Toward Sustainable Communities

Transition and Transformations in Environmental Policy

Edited by
Daniel A. Mazmanian
and
Michael E. Kraft

The MIT Press
Cambridge, Massachusetts
London, England

This book was set in Sabon by Wellington Graphics.

Printed and bound in the United States of America.

Library of Congress Cataloging-in-Publication Data

Toward sustainable communities : transition and transformations in
  environmental policy / edited by Daniel A. Mazmanian and Michael E.
  Kraft.
    p.   cm. — (American and comparative environmental policy)
  Includes bibliographical references and index.
  ISBN 0-262-13358-X (hc : alk. paper). — ISBN 0-262-63194-6 (pbk. :
  alk. paper)
  1. Environmental policy—United States. 2. Sustainable
development—United States. I. Mazmanian, Daniel A., 1945–   .
II. Kraft. Michael E. III. Series.
GE180. T69   1999
363.7′00973—dc21   99-34658
                                                              CIP

Toward Sustainable Communities

American and Comparative Environmental Policy
Sheldon Kamieniecki and Michael E. Kraft, editors

Critical Masses: Citizens, Nuclear Weapons Production, and Environmental Destruction in the United States and Russia
Russell J. Dalton, Paula Garb, Nicholas P. Lovrich, John C. Pierce, and John M. Whiteley

Toward Sustainable Communities: Transition and Transformations in Environmental Policy
edited by Daniel A. Mazmanian and Michael E. Kraft

Contents

Tables, Figures, and Boxes

Boxes

Foreword

The terms "sustainability," "sustainable societies," and "sustainable development" dominate the current debate over present and future environmental policy. Unfortunately little attention is being paid to the precise definition of these terms, and much confusion surrounds their use. Also complicating our understanding of the meaning of sustainability is whether we are focusing (or should focus) our attention on small, clearly defined jurisdictions at the local level, or whether it is indeed necessary to analyze and achieve sustainability at the national, regional, or global level. At the heart of the controversy, of course, is the extent to which communities (however defined) can have continued and vibrant economic development on the one hand, and environmental protection and a high quality of life on the other. What actions or policies actually constitute a move toward or away from developing a sustainable community depends upon the perspective of the analyst. Everyone now agrees that both goals are indeed attainable, but no one knows exactly how to balance the two and what decision rules and regulatory mechanisms to apply when the two aims run counter to each other.

This book addresses all of these questions concerning sustainability head-on by reviewing and assessing environmental policy successes and failures over the last three decades, primarily in the United States. The book deals at length with the various meanings of sustainability and its application in recent years at the local and regional level. Using six case studies of specific policy arenas, the editors and contributors examine the changing character of environmental policy within three overlapping but nonetheless distinct epochs. A main theme of the study is the movement of policy making away from command-and-control approaches and toward policy strategies that are associated with sustainability, particularly at the local and regional level. The book assesses the capacity of the concept of sustainability to serve as a foundation for a new era of environmental policy at all levels of government. As a consequence of evaluating the extent and pace of this overall transition to sustainability-based environmental policy, we come away with a much deeper understanding of the factors that affect the transition within different institutional and political contexts. Both policy makers and students of environmental policy will find this study provocative and stimulating, especially given the growing importance of the concept of sustainability.

Sheldon Kamieniecki, coeditor
American and Comparative Environmental Policy series

Preface

Anyone attempting to understand the state of environmental policy in the United States today will encounter a wide and bewildering array of laws, programs, and approaches. They range from very stringent and federally directed efforts, such as those in the air pollution and nuclear waste arenas, to an enormous assortment of local, regional, and state water, land-use, urban growth, and resource management policies. There are policies that are quite specific (for example, monitoring the transportation of hazardous waste), to the extremely broad (as with the congressional mandate to clean up the nation's waters to the point where they are swimmable and fishable). Policies differ in mission and purpose, in who is considered the cause of the problem, in who is responsible for remedying the situation, in what strategies are employed to bring about the desired change in behavior by individuals, businesses, and governmental agencies, and in who is expected to enforce the law.

The conventional way of bringing a degree of clarity to this confusing array is to trace the history of a particular type of environmental policy, such as water or air pollution. This is the most common approach in studies of environmental policy and is characteristic of most textbooks in the field. This "media"-specific approach is valuable and usually sheds light on the many and often unique details that exist within a given policy domain. The drawback to this approach is that it is never clear if the experience in one arena is unique or symptomatic of broader currents within the environmental movement and direction of environmental policy nationwide. An alternative and equally prominent approach is to provide a cross-sectional snapshot, spanning a range of environmental arenas. This is usually informative, but seldom does it lend itself to an understanding of the underlying trends.

We are persuaded that it is most important to understand the broad trends of environmental policy—which we refer to as the underlying environmental movement—and to do so in their boldest relief. Knowing the broad trends helps one fit the particulars from any specific arena into a more understandable overall picture. It is like having a good roadmap that while leaving out most of the details of the terrain, guides one forward in the journey—in this case, to an understanding of how environmental policy has unfolded and continues to unfold in an understandable way. This kind of conceptual clarity is especially necessary in the environmental arena, with its wide range of programs and strategies. Understanding the underlying trends also enables one to make meaningful comparisons; across policy domains—water, air, transportation, land management—and over time from one era to another. Last, knowing how to distinguish between the kinds of policies and activities that represent past efforts and current conventional thinking, and being able to distinguish these from leading-edge thinking and policy proposals, prepares one for active engagement in moving the environmental agenda and the environmental movement forward.

The fundamental thesis of this book is that the modern environmental movement can be best understood as the unfolding of three distinctly different but internally coherent epochs; the rise of environmental regulation, the period of flexibility and regulatory reform, and the move to sustainable development. This movement begins with the national awakening to the problems of environmental pollution and issues this raised in the late 1960s, and moves through to the merging with the broader movement toward sustainability, which extends well beyond the boundaries of environmental policy and the concerns of the first two environmental epochs. Hence, what at first blush might appear to be a chaotic mix of goals, approaches, programs, and activities when viewed today can be understood as the cumulative consequence of two past epochs and a third emerging epoch of environmentalism. Furthermore, when viewed separately, each epoch has a fairly coherent internal logic and policy agenda, and understanding environmental policy in the context of each helps distinguish the broad contours of environmentalism. The epochs framework provides a valuable roadmap to the historical evolution, logic, and policy actions of the movement.

The most controversial issue in the environmental movement today involves the likelihood of its moving from the goals, strategies, and approaches to protecting and restoring the environment developed in its first two epochs to those needed to achieve a more sustainable society, and by doing so transforming the movement into the third modern epoch. Although the rhetoric of sustainability is being espoused widely and has enormous personal and intellectual appeal to environmentalists (and many business and political leaders as well, all the way up to the president of the United States), the reality of the transformation is less obvious. Indeed, little scholarly and systematic evidence exists on the extent to which the rhetoric is being put into practice, or if it ever can be.

Although theorists and pundits will continue to debate the issue, the world of practice is not standing still. Environmental pollution and degradation continue to be serious problems, and people at the local and regional level, persuaded that sustainability is the only viable path, are taking matters into their own hands. What is needed at this point therefore is a clearer matching up of the theoretical virtues of more sustainable systems with the experiences of sustainability initiatives in the field.

One of the main purposes of the book is to provide a synthesis of theory and practice by tracing the environmental movement through two critical transformations. These are the transformation from the first epoch, characterized by an array of federal environmental laws and top-down federal regulation, to the second, where emphasis is placed on incentive-based policy approaches and community and regionally based decision making. This is illustrated through initiatives under way within air, water, and land-use policies.

We then move from the traditional environmental concerns to the more comprehensive and complex systems approach that takes into consideration many more factors affecting a community's overall sustainability. Doing so requires moving beyond conventional environmental issues to incorporating a community's physical, psychological, economic, and cultural well-being. The degree of success of these pilot efforts—in urban redevelopment, intermodal transportation, and region-wide environmental management—will be extremely valuable not only for what they suggest about the future direction of the environmental movement, but also for American society as a whole at the start of a new century. We find the experiences highly informative and promising of a more sustainable environmental—and social, economic, and cultural—future.

We thank the contributing authors for their cooperation and diligence in response to the several rounds of editorial requests that we and MIT Press made. It has been a pleasure to work with such a dedicated and talented group of colleagues. We also appreciate the detailed comments on the draft manuscript by our anonymous reviewers, who helped to improve each chapter and develop a more coherent book. We gratefully acknowledge support from the School of Natural Resources and Environment at the University of Michigan and the Department of Public and Environmental Affairs at the University of Wisconsin-Green Bay. As always, we take full responsibility for any remaining errors or omissions.

Contributors

Daniel A. Mazmanian is Dean of School of the Natural Resources and Environment and Professor at the University of Michigan.

Michael E. Kraft is Professor of Political Science and Public Affairs and Herbert Fisk Johnson Professor of Environmental Studies at the University of Wisconsin-Green Bay.

Lamont C. Hempel is Hedco Chair of Environmental Studies and Director of Environmental Programs at the University of Redlands, Redlands, California.

Bruce N. Johnson is Executive Director of the Fox-Wolf Basin 2000 in Northeast Wisconsin.

Daniel Press is Associate Professor in the Environmental Studies Department at the University of California at Santa Cruz.

Franklin Tugwell is President and CEO of Winrock International Institute for Agricultural Development and past Executive Director of the Heinz Endowments in Pittsburgh, Pennsylvania.

Andrew S. McElwaine is President of the Pennsylvania Environmental Council and past Director of Environmental Programs at the Heinz Endowments.

Michele Kanche Fetting is Communications and Research Consultant at the Heinz Endowments.

Thomas A. Horan is Director of the Claremont Graduate Research Institute at Claremont Graduate University, Claremont, California.

Hank Dittmar is Director of the Quality of Life Campaign and former Executive Director at the Surface Transportation Policy Project in Washington, D.C.

Daniel R. Jordan is a Ph.D. candidate at Claremont Graduate University and consultant at the firm of Hamilton, Rabinovitz, and Alschuler, Inc., Los Angeles.

Barry G. Rabe is Professor in the School of Natural Resources and Environment at the University of Michigan.

I

Introduction

1

The Three Epochs of the Environmental Movement

Daniel A. Mazmanian and Michael E. Kraft

The ability of a country to follow sustainable development paths is determined to a large extent by the capacity of its people and institutions as well as by its ecological and geographical conditions. Specifically, capacity-building encompasses the country's human, scientific, technological, organizational, institutional and resource capabilities. (United Nations, *Agenda 21*, Ch. 37, 270)

In the short three decades since the onset of the modern environmental movement, an enormous amount has been learned about the capacity and limits of government's ability to direct economic activity, affect human values and behavior, and create a more livable and sustainable world. The United States has been one of the important crucibles of this learning process, with the resultant widespread awareness of the delicate interrelationships among social, economic, and environmental systems. A great deal has been accomplished, as well, in appreciably reducing air, water, soil, and other pollutants in the environment (Portney 1990a; Council on Environmental Quality 1997; Kraft 1996; Davies and Mazurek 1998). Despite these strides, not all is well with the environment, environmental policy, and conceptual approaches employed to understand and interpret what is taking place in this most important revolutionary movement of the age. We are at a crossroads in several senses of the word and need to take stock and reconsider the future.

About the environment itself, we have learned how to accomplish significant reductions in air, water, and land pollution, both per capita and overall. The evidence here is irrefutable (Davies and Mazurek 1998, ch. 5). Yet it is also clear that these gains *may* be short-lived as greater amounts of fossil fuel energy and materials are consumed, new threats to the environment become known—such as the buildup of greenhouse gases, deterioration of the stratospheric ozone layer, and emission of very minute particulates in the air—and the stock of species and nonrenewable natural resources is depleted (Janicke, Monch, and Binder 1993). Also, curbing harmful development and human expansion in one place—a pristine coastline, wetlands, a unique landscape, or an endangered species habitat—does not prevent it from surfacing somewhere else. The extensive effort of the federal government to clean the nation's environment has come at times at the price of conventional economic development for business and industry, and it has contributed to the migration of some industries to other countries. Developing and implementing the nation's environmental laws and regulations also has led to the creation of substantial government bureaucracies, which can frustrate as much as help remedy environmental problems (Landy, Roberts, and Thomas 1994; Rosenbaum 1997).

Moreover, it does not appear from today's vantage point that continuation of the stringent regulatory approach initiated in the early 1970s is either an appropriate or feasible strategy for realizing the long-term goals of environmental sustainability (Fiorino 1996; National Commission on the Environment 1993). The lesson of many years of work in the field leads back to one of the oldest adages of the environmental movement, that is, to think and plan globally but act locally (Mazmanian and Morell 1992). Doing so requires mobilizing the capabilities of local communities. Only by doing this can the issues of the size and scale, together with the human dimensions and unique cultures and settings, be adequately molded into enduring and sustainable environmental strategies.

Today, therefore, the conventional policy framework of environmental protection is undergoing a reassessment on three basic fronts. One is its overreliance on command-and-control regulation, which despite notable achievements over the past three decades cannot be the only or perhaps even the major strategy for achieving environmental sustainability (Stavins 1991; John 1994). Among the most important limitations of conventional regulation are its high costs and inefficiencies, a focus on remedial rather than preventive actions, and complex, cumbersome, and adversarial rule-making processes (Meiners and Yandle 1993). These weaknesses have become especially significant at a time of stagnant or diminishing budgetary resources and intense opposition from affected industries and state and local governments—problems exacerbated by widespread public distrust of politicians, distant government agencies, and social regulation, and a failure of democratic dialogue in the nation (Williams and Matheny 1995).

Second, national environmental policy often leaves both industry and state and local governments without needed flexibility and incentives for achieving environmental quality objectives. Critics argue that the federal Environmental Protection Agency (EPA) often has proven to be inept at priority setting and program management (Landy, Roberts, and Thomas 1994; National Academy of Public Administration 1995). Centralized policy dominance may be necessary in areas where the states fall well short of the minimal standards necessary for protection of public and ecological health. However, in an age when state and local governments are more capable of policy action and willing to pursue environmental quality goals (Davies and Mazurek 1998, ch. 4), greater attention needs to be paid to the forces encouraging or inhibiting policy innovation at the subnational level and the impacts of such innovation on environmental quality (Hamilton 1990; Mazmanian and Morell 1991; Lowry 1992; Ringquist 1993; Lester 1995; Rabe 1997).

The third limitation of the conventional approach is, ironically, its relative neglect of sustainable development. Somehow this end was lost sight of in pursuit of statutorily prescribed means. In sectors as diverse as air and water pollution, energy, agriculture, construction, transportation, land use, and urban planning, economic and environmental goals must be reconciled, integrated, and directed toward achievement of sustainable communities. This task is beyond the reach of today's national environmental protection policies. Those policies must be rewoven into a holistic program of environmental protection and the pursuit of improvements in the quality of life for human populations, from the local through the global level (Trzyna 1995).

National policy leaders have begun to recognize the imperative of sustainability, as witnessed by the recent reports of President Clinton's Council on Sustainable Development (1996) and pollution prevention and sustainable community initiatives of the Environmental Protection Agency and the departments of Housing and Urban Development and Energy during the past several years (National Commission on the

Environment 1993; Press and Mazmanian 1997; Sitarz 1998; U.S. EPA 1997). Many of the most promising sustainability efforts, less noticed, perhaps, but with enormous potential, can be found in the growing application of new approaches at the state, local, and regional levels across the nation (Hren, Bartolomeo, and Signer 1995; Minard 1996; Hempel 1998; Sexton et al. 1999) and around the globe (Bartone et al. 1994; Trzyna 1995; Maser 1997).

Innovative actions to foster sustainable development worldwide were at center stage at the 1992 United Nations Conference on Environment and Development (the Earth Summit) and highlighted in its detailed program of action, Agenda 21 (United Nations 1993; Vig and Axelrod 1999). Despite the many obstacles to instituting such policies and programs, the 1990s have witnessed a surge of interest in the promise of sustainable communities, as well as a diversity of intriguing developments, particularly in Western Europe (e.g., Nijkamp and Perrels 1994; Liefferink 1999; Axelrod and Vig 1999). We consider these efforts, especially the adoption of new approaches to environmental protection, as beacons for the future and thus make them the primary focus of our attention.

As noted, criticism of environmental policy as being too costly, bureaucratic, narrow, and over-zealously pursued has become commonplace. Yet such blanket indictments help little in identifying the roots of policy failure or directions for the future. It is remarkable that after thirty years of modern environmental policy we have so little systematic evaluation of the dozens of policies and hundreds of programs being overseen by the EPA and the fifty states (Knaap and Kim 1998). It is also clear that proposed alternatives have remained largely unexamined, although they often have great theoretical appeal. This has been no different for those aspiring to sustainable communities. While they have made persuasive appeals and have gained prominence among environmental and some community leaders, their ideas have remained mostly untested to date. Why this is the case will become more evident in chapter 2.

Nevertheless, what is being called for is a transformation from conventional policies and understandings about how best to protect the environment as they have evolved over the past three decades to ideas believed to be more sustainable. The primary objective of this book is to help readers understand and envision this transformation. We begin with some of the more fundamental changes that have been under way for a decade or more in the traditional regulatory approach to environmental protection in air, water, and to some extent in land use (this is the transition from the first to the second environmental epoch, as we will describe later). While not end points in themselves, these changes reflect an important step beyond the dominant, centrally managed regulatory approach set in place in the early days of the modern environmental era. We characterize them as bridging to a more sustainable approach in environmental protection (epoch three, as discussed later). Following this assessment are chapters that focus on the more recent generation of policy approaches and commmunity and regional experiments in sustainability that point the way to the future, raising the specter of a second fundamental transformation in the modern environmental movement to come.

Organizational and Conceptual Overview

Focus on Environmental Epochs

Part of the difficulty in making sense of the environmental movement today and its implications for society's future is that there are almost as many ways of depicting the movement as there are approaches to understanding nature and society in the natural and social sciences, humanities, and policy professions. The causes and consequences of—and the remedies sought for—environmental problems follow from each disciplinary framework or paradigm that points to some critical driving force in society, where best to intervene, and the kinds of public policies that are preferable.[1] As useful as disciplinary based explanations are in helping to frame the issues of environmental protection in familiar terms for different academic and professional audiences, the approach is too narrow.

What history tells us is that solutions to most environmental problems have not resulted from a specific approach but have required input from a multiplicity of perspectives. They have evolved through an organic process of trial, error, and societal learning. Also, it is clear in retrospect that there has been a progression in the way people have framed and dealt with environmental issues. To make sense of the present while trying to anticipate the future, it is important to understand this progression. The progression, which at one level has been quite incremental, overall has occurred in a small number of distinct epochs. Each epoch is characterized by a dominant way of defining "the" environmental problem (comprised of both a scientific and value component), which in turn leads to a set of policy goals, the use of certain implementation strategies, and other features that must be considered together to capture the essence of the epoch.

Understanding the historical sequence of epochs is important also in that each builds on that which preceded it, ultimately overshadowing (in terms of dominant ideas and focus) and overlaying it (in terms of policies and programs) but never fully replacing it—along with all the confusion and complexity such progression leads to (Mazmanian and Morell 1988). Like a good map, the epochs approach attempts to outline the key features of the landscape and show the links between past and present, on the one hand, while indicating, on the other, how each is distinct in some fairly fundamental ways. The focus on epochs also enables us to stand back from the details and narrow views that come with everyday life and grasp the overall features of the environmental movement at each major juncture in its history.

Finally, there has been a dramatic growth in understanding about the environment over the past three decades, with each epoch bringing more into focus the interdependence of human and natural systems and the ultimate limitation of the Earth's ability to sustain infinitely expanding human populations and levels of material consumption, development of technological solutions, and considerations of justice and equity. Understanding the "map" of the first two epochs of the modern environmental movement, combined with the growing awareness of the threats to the health of the natural environment at home and around the world, is the basis for our forecast (in effect, our best guess). We think it is both necessary and likely that the nation will move to a more enduring and sustainable epoch in which concerns for the natural environment will play a far more pronounced role. We also expect that the transition will occur at widely varying rates and in different forms from one region of the nation to another and across communities.

Table 1.1 presents the three epochs around which the book is organized, beginning in the early 1970s with the rise of environmentalism as a social and political movement and the buildup of the system of federal command-and-control environmental regulation, with its hallmark clean air and clean water legislation, creation of the U.S. EPA, and strong federal presence. The third epoch brings into focus the potentiality of sustainable development and sustainable communities, looking to the twenty-first century. The second epoch in between is transitional in several key respects. It is marked by the drive for efficiency and flexibility in the regulatory apparatus created in the first epoch. Its rhetoric and politics are dominated by those with business and property holdings who saw themselves adversely affected by the new generation of environmental laws in the first epoch. Future historians will likely characterize this second epoch as a bridging one. The table provides an overview and highlights the critical dimensions of the three epochs and major differences among them in problem identification and policy objectives, implementation philosophies, points of intervention, policy "tools," data and informational needs, political and institutional contexts, and key events and public actions. We believe these features define and differentiate the epochs from one another, and in combination gives each its overall meaning.

This book seeks to present the map of each epoch and to explore how useful the mapping framework is in illuminating the unfolding of the environmental movement. To do this, we have asked several prominent environmental policy scholars and keen observers to contribute, bringing to bear their knowledge of a community or policy arena to test how the epochs approach fits. They will determine how well the epochs approach provides a basis of understanding through identifying the analytical dimensions around which their cases can be explained.

Problem Definition and Policy Objectives

The objective of the first modern environmental epoch was to place center stage the necessity of cleaning up the nation's polluted waterways, air,

Table 1.1
From environmental protection to sustainable communities

	Regulating for environmental protection, 1970–1990	Efficiency-based regulatory reform and flexibility, 1980–1990s	Toward sustainable communities, 1990-on
Problem identification and policy objectives	• pollution caused primarily by callous and unthinking business and industry • establish as national priority the curtailment of air, water, and land pollution caused by industry and other human activity	• managing pollution through market-based and collaborative mechanisms • subject environmental regulations to cost-effectiveness test • internalize pollution costs • pursue economically optimal use of resources and energy • introduce pollution prevention • add policies on toxic waste and chemicals as national priorities	• bringing into harmony human and natural systems on a sustainable basis • balance long-term societal and natural system needs through system design and management • rediscover/emphasis on resource conservation • halt diminution of biodiversity • embrace an eco-centric ethic
Implementation philosophy	• develop the administrative and regulatory legal infrastructure to ensure compliance with federal and state regulations	• shift to state and local level for initiative in compliance and enforcement	• develop new mechanisms and institutions that balance the needs of human and natural systems, both within the U.S. and around the globe

Points of intervention	• end of the production pipeline • end of the waste stream • at the point of local, state, and federal governmental activity	• create market mechanisms for protection of the environment • the market-place, which serves as the arbiter of product viability • provide education and training at several points along the cradle-to-grave path of materials and resource use	• societal-level needs assessment and goal prioritization • industry-level attention to product design, materials selection, and environmental strategic planning • individual behavior and lifestyle choices
Policy approaches and "tools"	• policy managed by Washington, D.C. • command-and-control regulation • substantial federal technology R&D • generous federal funding of health and pollution prevention projects	• policy managed more by states and affected communities • federal role shifts to facilitation and oversight • introduction of incentive-based approaches (taxes, fees, emissions trading) for business and industry • creation of emissions-trading markets	• comprehensive future visioning • regional planning based on sustainability guidelines • Total Quality Environmental (TQEM) and life-cycle analysis practice in industry • various experiments with new approaches

Table 1.1
(cont.)

Information and data management needs	• company-level emissions • waste stream contents and tracking • human health effects • environmental compliance accounting in industry	• costing out environmental harms and benefits of reduced pollution • provision of readily accessible emissions data, e.g., through Toxics Release Inventory and right-to-know programs • professional protocols for environmental accounting in industry • ecosystem mapping	• sustainability criteria and indicators • eco-human support system thresholds • region/community-global interaction effects (e.g., regarding CO_2 emissions and depletion of ozone layer) • utilization of ecological footprint analysis • use of material and energy "flow-through" inventories and accounting • computer modeling of human-natural systems interactions
Predominant political/institutional context	• rule of law • adversarial relations • zero-sum politics	• alternative dispute resolution techniques • greater stakeholder and public participation, especially at the state and local levels • reliance on the market place	• public/private partnerships • local/regional collaborations • community capacity building and consensus building

	Epoch 1	Epoch 2	Epoch 3
	• focus on national regulatory agencies and enforcement mechanisms		• mechanisms created to enforce "collective" decisions
Key events and public actions	• Santa Barbara oil spill • Earth Day • passage of the 1970 CAA and 1972 CWA • passage of National Environmental Policy Act • creation of the Environmental Protection Agency	• Carter administration focus on cost of environmental regulation • election of President Ronald Reagan • Love Canal, Bhopal • RCRA and SARA • growth in state and local environmental policy capacity	• attention to global issues of sustainability • Brundtland report, *Our Common Future* • Earth Summit (UNCED) • collective international action—Montreal Protocol on CFCs, international accords on global warming

and land. Which business and industrial activities were responsible for the pollution was another matter, and subject to a great deal of debate. For instance, were automobiles, industrial facilities, or climatological conditions the major source of urban air pollution? Whatever the cause, the solutions proposed were almost always costly and therefore contentious. Yet during the first epoch a consensus emerged among scientists, technicians, policy makers, and the public that the issues of pollution and environmental degradation were severe and should be addressed as a top national priority (Dunlap 1995; Bosso 1999). Despite the criticism that would eventually be voiced about the cleanup effort these decisions precipitated, there is little question that the first environmental epoch produced significant improvements in air and water quality in the United States and made important gains in reducing the careless disposal of hazardous wastes and toxic chemicals (Portney 1990a; Council on Environmental Quality 1997; U.S. EPA 1998a and 1998b).

Beyond those policies directed at pollution control, implementation of the National Environmental Policy Act of 1969, with its broad mandate for comprehensive impact assessment and public involvement in environmental policy decisions, spurred significant changes across federal and state bureaucracies (Caldwell 1982). Protection of the nation's natural resources was advanced substantially during this era through adoption of new policies with extensive federal mandates for protection of biological diversity and for the stewardship of public lands through what would later be called ecosystem management. These include the Endangered Species Act (1973), the Federal Land Policy and Management Act (1976), and the National Forest Management Act (1976), among others (Kraft 1996; Cortner and Moote 1998; Yaffee et al. 1996).

In the second epoch, the focus shifted to balancing environmental objectives with other social and economic priorities and to carrying out more efficiently those environmental policies that were on the books. In a few instances goals were expanded, such as adding toxic materials and hazardous waste to the environmental policy agenda, the more demanding provisions of the Clean Air Act of 1990, and the greater recognition of the international and global ramifications of pollution (Mazmanian and Morell 1992; Bryner 1995; Hempel 1996). Overall, the pace of legislation and coverage of newly identified sources of pollution slowed appreciably in comparison with the first epoch.

What changed most noticeably was faith in the philosophy of regulation and strong control by the federal government. It became clear during the first epoch that government alone, especially the federal government, could neither direct nor police all businesses and every community across the nation, nor could it shoulder all the responsibility for stimulating innovative responses to environmental problems. This was not simply a reaction against an ever-growing government involvement. Underlying the second epoch was the recognition that the corner had been turned in many areas of environmental pollution and resource protection. Also, after more than a decade of attention, problems of the environment were no longer seen as quite as catastrophic as the media had portrayed them throughout the first epoch (although the cumulative and chronic implications of pollution and pressures on natural resources remained). Concerns would hence be weighed in the balance against competing social needs and the pressures for (near-term) economic development. These shifts occurred within the broader framework of the growing conservative and anti–federal government tide that swept the country in the latter part of the 1970s and into the 1980s.

One result was that environmental regulations and proposed solutions would be required to submit to conventional cost-effectiveness tests (Portney 1990a; Meiners and Yandle 1993; Freeman 1997) and risk assessments (U.S. EPA 1990; National Research Council 1994; Davies 1996). In addition, the locus of decision making would devolve in greater measure from federal actors to those in affected state and local communities.

The atmosphere of reassessment, regulatory devolution and flexibility, and cost-consciousness provided an opening for pent-up political and business opposition to environmentalism, not only to modify policies but to alter basic objectives. The backlash can be bracketed by the effort (partially successful) to slash the environmental protection budgets of the federal government during President Ronald Reagan's first term in the early 1980s, and to the watershed election of 1994, in which environmental policies were a prime target of the new conservative Republicans' "Contract with America" (Vig and Kraft 1997). While this revealed a fundamental change in goals, the actual effect was less the sought after reversal in policy than the dampening of legislative zeal and policy expansion.

The lessons of the first two epochs were not lost on those concerned with the health of the population and the natural environment. Improvements could be made for waterways, air sheds, and waste sites across the land through strong, forceful, and aggressively enforced federal and state environmental laws. Yet serious challenges would remain and new ones continue to emerge—such as the loss of biological diversity, the need for habitat management and open space, the possible adverse effects of global warming, and the possibility of a population growth of 50 percent, to some 390 million people in the United States, over the next six decades.

The close linkage between human population growth, settlement patterns, and industrial activity and the degradation of the environment could not be ignored if permanent solutions were to be found. These relationships were underscored first in developing countries, where several cases of resource depletion and industrial pollution had become severe seemingly overnight (Trzyna 1995). But eventually they were seen as the root causes of much of America's environmental, health, social, and natural resources issues as well. Problems of the environment were neither simple to address nor isolated from the pace and growth of other human activities, and they could be remedied only with sustained, comprehensive, multigenerational efforts.

The decision by a growing number of people from all walks of life to address the transformational needs of society is the hallmark of the third environmental epoch. Going beyond prevention and restoration, and determining the most cost-effective methods for doing so pales in comparison to the goal of sustainability. Focusing on sustainability, for instance, draws attention to the failure to incorporate into the economic activity in society—and calculation of a nation's gross national product— measures of environmental health, quality of life, and the full (true) costs of human settlement patterns on the land and the consumption of natural resources (World Commission on Environment and Development 1987; Cobb, Halstead, and Rowe 1995; Daly and Cobb 1989; Daly 1977; Van Dieren 1995).

Discussions are under way on how best to include these considerations, as chapter 2 will detail. For example, Cobb and others, in developing their measure of "genuine societal progress," conclude that a more complete national accounting would reveal a downward trend in the genuine per capita level of wealth of all Americans since the mid-1970s. Such assertions are hotly debated in conventional economic circles. Nonetheless, the inclusion of the depletion of natural resources, the costs of environmental degradation, and other negative societal costs in some aggregate national measure reflects the latest thinking by today's leading environmentalists. Increasingly, public policy makers recognize the appropriateness of this approach.

The efforts to transform the way we account for the nation's wealth only begs the broader question of how so intuitively appealing yet vague an idea as sustainability is to be defined—and there is no simple answer. For some, sustainability is a model, "an ideal set of goals to work toward. But it is also a philosophy for envisioning those goals and a practical problem-solving process for achieving them" (Geis and Kutzmark 1995). It is gradually becoming an ethical standard for humans to live by (Milbrath 1989; Westra 1994), and, in turn, a set of principles by which communities, commerce, and industry are being judged (Schmidheiny 1992; Smart 1992; Hempel 1996 and 1998; Press and Mazmanian 1997). The simplest and possibly most encompassing definition of sustainability was provided by the World Commission on Environment and Development (Brundtland Commission) Report (1987, 43): "meeting the needs of the present without compromising the ability of future generations to meet their own needs." What exactly constitutes needs and how to meet them are questions that remain open.

One important distinction being made is between "weak" and "strong" sustainability. In the weak variety, the present generation has an obligation to pass on to future generations an average capital stock—of goods, services, knowledge, raw materials—that is equivalent to today's. In essence, taking all natural and human resources together, the current generation is obliged not to deplete the total stock. While any given generation may deplete certain resources, as long as those can be replaced through human invention, the process is sustaining. The "strong sustainability" school, in contrast, sees certain natural stocks as essential ecological resources and building blocks for the much broader ecosystem (e.g., the ozone layer and biodiversity), thus inappropriate for averaging in with other kinds of assets (e.g., energy-efficient and low- polluting technologies). Not all assets are the same and, for the strong sustainability school, some natural resources and ecological processes are critical; they cannot be depleted below a certain level without dramatic ramifications for sustainability. Thus they cannot be easily averaged into an intergenerational balance sheet (Van Dieren 1995).

As a practical matter, sustainability can mean any important change in values, public policy, and public and private activity that moves communities and individuals toward realization of the key tenets of ecological integrity, social harmony, and political participation (Hempel 1996 and 1998). In the United States, activities that qualify fall under several headings, in addition to sustainable communities, including "urban ecology," "sustainable development," "sustainability planning," and "greening," to name a few.

There is obviously a great deal of ambiguity today in the concept of sustainability, and related ones such as the "carrying capacity" of the planet. Moreover, predicting future rates of economic growth, consumption, and pollution remain highly complex and problematic. Nevertheless, there is growing recognition that human populations cannot expand indefinitely given the physical limitations of the earth's land mass and resource base and human dependence on critical ecological processes (Ophuls and Boyan 1992; World Commission on Environment and Development 1987). It is possible to imagine a trade-off between the absolute size of the planet's population—or that of a town or community—and its energy and resources support systems. A population that consumes less per capita can sustain a larger size over time.[2] For every combination of population size and average resource use, however, there exists an absolute limit beyond which the capacity of ecosystems to sustain human beings breaks down (Gray 1993, ch. 14). Yet this portrayal of the relationship leaves open the question of how much and what kind of economic growth is tolerable, and within what time frame the limits will be reached. Determining where these thresholds lie is one of the central questions for analysis for the third epoch of environmentalism (Wackernagel and Rees 1996), as will be clear in the next chapter.

Implementation Philosophy, Points of Intervention, and Policy Tools

Implementation philosophy goes to the heart of beliefs about how best to achieve agreed upon public policy goals (Mazmanian and Sabatier 1989), and this heavily influences the points of intervention selected and policy tools adopted. Even when different groups and officials can agree on what they want accomplished, determining how best to do so may not be easy. Should people be coaxed or compelled to act a certain way? Should noncompliance be punished, and if so, how severely? Should emphasis be placed on educating people and providing them the where-withal to change, or should they be expected to change behavior, irrespective of costs or their level of awareness of alternatives, as a matter of law? Furthermore, the status, power, and public perception of the groups the legislation is intended to affect often have a great deal to do with the implementation philosophy adopted by political leaders and, in turn, what policy tools are utilized and where (Schon and Rein 1994; Schneider and Ingram 1990 and 1997).

Seldom explicit, implementation philosophy is usually embedded in the mechanisms Congress, state legislatures, and communities establish to carry out public policies. Their understanding of the problem and choices of how best to bring about the desired changes in people's actions are revealed in how they decide to assign various responsibilities. For example, they may assign a task to an existing federal, state, or local agency. Or they may create a new agency for the job, or assign it to an existing regulatory commission, or even to a variety of public/private or even wholly private organizations. They may decide to criminalize certain kinds of behavior—such as disposing of hazardous waste on land—and to invoke major penalties for violations, or make them minor violations with minimal penalties.

The implementation philosophy of the first environmental epoch was long on process and building new governing institutions, along with oversight of government activities as they affected the environment, but short on actually dictating the behavior of business, industry, and individuals. The signals were clearly mixed, but a combination of both "stick" and "carrot" was utilized.

Probably the most important feature of the first epoch's philosophy was that policy needed to be centralized in the hands of a new comprehensive federal agency: the U.S. Environmental Protection Agency. Given the level of state policy capacity at the time and the failure of most states to aggressively pursue protection of even their own environments, it was widely believed that if the nation's air, water, land, and related pollution problems were to be addressed successfully, it would have to be done under strong national, uniform guidelines and enforcement by a single agency, along with forceful legislation in critical areas of concern. The most important "seven pillars" of environmental protection legislation from this era are highlighted in box 1.1. For this purpose we exclude the equally important natural resource policies adopted at about the same time, such as the National Environmental Policy Act of 1969. This core of environmental protection or pollution control statutes was developed at a time when it was believed that a "big stick" was necessary to bring about change, and that state and local governments either could not or would not be forceful enough. This top-down approach can be contrasted with the decentralized, though still governmental, approach of the second epoch, and the community-based, more integrative strategy envisioned for the third epoch.

The administrative task this approach presented to the EPA was formidable, particularly for a regulatory bureaucracy struggling to gain legitimacy and sufficient resources and do its job while fending off its critics, both from outside and within the federal government. Moreover, what policy and government capacity did exist to deal with air, water, and land pollution was spread among different agencies with little history of coordinated action. In recognizing this history and the inherent difficulty of developing a more integrated approach, the EPA simply extended and reinforced a pollution or problem-specific organizational structure, which persists to the present, with separate program offices for air and radiation, water, pesticides and toxic substances, and solid wastes and emergency response.

To carry out its expanding mission, the agency's staff grew from about 7,000 in the first full year of operation to about 18,000 by the mid-1990s, some two-thirds of the employees working in the agency's ten regional offices and in other facilities located outside of Washington, D.C. The agency's operating budget rose from an initial $500 million in 1971, to about $3 billion by 1998. However, adjusted for inflation, the operating budget in 1998 was scarcely higher than was it was in the mid-1970s. A

Box 1.1
Seven pillars of the first environmental epoch

1. The Clean Air Act (CAA). The 1970 act required the EPA to set uniform, national ambient air quality standards to "provide an adequate margin of safety" to protect public health "from any known or anticipated adverse effects" associated with six major pollutants. Enforcement was to be shared by the federal government and the states. The states were to develop implementation plans to achieve compliance with the new air quality standards. The act also set national emissions standards for mobile sources of air pollution (cars, buses, and trucks) and standards for stationary sources such as refineries, chemical companies, and other industrial facilities. In the 1990 revision of the act, Congress created a market-based emissions trading program for control of sulfur dioxide and nitrogen oxides that contribute to acid precipitation, set more stringent automobile emission standards, required the EPA to set new emission limits to control all major industrial sources of hazardous or toxic air pollutants, and set out an elaborate multitiered plan to bring all urban areas into compliance with national air quality standards in three to twenty years.

2. The Clean Water Act (CWA). Formally the Federal Water Pollution Control Act Amendments of 1972, the CWA set a national policy for cleaning up the nation's surface water. It established national deadlines for eliminating discharge of pollutants into navigable waters by 1985, and set as a goal "fishable and swimmable" waters nationwide by 1983. Primary responsibility for implementation is given to the states as long as they follow federal standards and guidelines. Both industry and municipal dischargers must apply for permits to discharge pollutants, which are governed by applicable water quality criteria. Economic costs of pollution control may be considered as part of the standard setting and permitting process. An initial federal grant program for construction of municipal sewage treatment plants was redesigned in a 1987 amendment as a revolving loan fund to provide seed money for wastewater treatment. The 1987 act also required states to develop EPA-approved plans for control of nonpoint sources of water pollution.

3. The Safe Drinking Water Act (SDWA). The 1974 act was designed to ensure the quality and safety of drinking water by specifying minimum public health standards for public water supplies. It authorized the EPA to set National Primary Drinking Water Standards for chemical and microbiological contaminants in tap water. The act also required regular monitoring of water supplies to ensure that pollutants stayed below safe levels. The 1986 amendments required the EPA to determine maximum contaminant levels for eighty-three specific chemicals by 1989, and set quality standards for them, set standards for another twenty-five contaminants by 1991, and twenty-five more every three years. In 1996, Congress dropped

Box 1.1
(cont.)

the last requirement and provided greater flexibility for the states in meeting federal drinking water standards in exchange for creation of a new
"right-to-know" policy. Communities are now required to notify consumers of the safety of local water supplies, and to publish information on a
broad array of detectable contaminants found in drinking water.

4. The Resource Conservation and Recovery Act (RCRA). In the 1976 act,
Congress required the EPA to regulate existing hazardous waste disposal
practices as well as to promote the conservation and recovery of resources
through comprehensive management of solid waste. RCRA required the
EPA to develop criteria for safe disposal of solid waste and the Commerce
Department to promote waste recovery technologies and waste conservation. The EPA was to develop a "cradle-to-grave" system of regulation that
would monitor and control the production, storage, transportation, and
disposal of wastes considered hazardous, and it was to determine the
appropriate technology for disposal of wastes. In the 1984 rewrite of the
act, Congress sought to phase out disposal of most hazardous wastes in
landfills by establishing demanding standards of safety; expanding control
to cover additional sources and wastes (particularly from small sources
previously exempt); extending RCRA regulation to underground storage
tanks (USTs) holding petroleum, pesticides, solvents, and gasoline; and
moving more quickly toward policy goals by setting out a highly specific
timetable for mandated actions.

5. The Toxic Substances Control Act (TSCA). In this 1976 act, the EPA
was given comprehensive authority to identify, evaluate, and regulate risks
associated with the full life cycle of commercial chemicals, both those
already in commerce as well as new ones in preparation. TSCA aspired to
develop adequate data on the effect of chemical substances on health and
the environment and to regulate those chemicals posing an "unreasonable
risk of injury to health or the environment," without unduly burdening
industry and impeding technological innovation. The EPA was to produce
an inventory of chemicals in commercial production, and it was given
authority to require testing by industry where data are insufficient and the
chemical may present an unacceptable risk. However, exercise of that
authority was made difficult and time consuming. Although the meaning
of "unreasonable risk" is not formally defined in the act, Congress clearly
intended some kind of balancing of the risks and the benefits to society of
the chemicals in question. TSCA was modified in 1986 to include actions
on asbestos and lead.

6. The Federal Insecticide, Fungicide, and Rodenticide Act (FIFRA). Congress created FIFRA in a 1947 act that established a registration and
labeling program housed in the Department of Agriculture that was ori-

Box 1.1
(cont.)

ented largely to the efficacy of pesticides. In 1972, Congress established the modern regulatory framework that turned jurisdiction over to the EPA. FIFRA requires that pesticides used commercially within the United States be registered by the EPA. It sets as a criterion for registration that the pesticide not pose "any unreasonable risk to man or the environment, taking into account the economic, social, and environmental costs and benefits of the use." Procedures under the law were cumbersome, making regulatory action difficult. In 1996, Congress approved legislation that requires the EPA to establish a level of exposure that ensures a "reasonable certainty of no harm" from pesticide residues in food, whether processed or raw agricultural products. The EPA will also publish pamphlets that will summarize the risks and benefits of pesticides and alert consumers to foods that have a high pesticide residue level.

7. The Comprehensive Environmental Response, Compensation, and Liability Act (CERCLA or Superfund). Congress enacted CERCLA, better known as Superfund, in 1980, and revised it in 1986, with the Superfund Amendments and Reauthorization Act (SARA). CERCLA is directed at the nation's thousands of abandoned and uncontrolled hazardous waste sites. Congress gave the EPA responsibility to "respond" to the problem by identifying, assessing, and cleaning up those sites. The EPA could use, where necessary, a special revolving fund of $1.6 billion, later increased to $10 billion, most of which was to be financed by a tax on manufacturers of petrochemical feedstocks and other organic chemicals and crude oil importers. The act put responsibility for the cleanup and financial liability on those who disposed of hazardous wastes at the site—a "polluter pays" policy. SARA also established a new Title III in the act, also called the Emergency Planning and Community Right-to-Know Act (EPCRA). It provided for public release of information about chemicals made by, stored in, and released by local businesses (published each year as the Toxics Release Inventory).

Sources: Michael E. Kraft, *Environmental Policy and Politics*, chap. 4 (New York: HarperCollins, 1996), and Norman J. Vig and Michael E. Kraft, eds., *Environmental Policy: New Directions for the Twenty-first Century* (Washington, DC: CQ Press, 1999).

lack of sufficient resources clearly has constrained the agency's capacity to keep up with the increased responsibilities that Congress has thrust on it over the years.[3]

Finally, it was assumed that the demanding new controls over environmental pollution could be put in place without substantially altering the affairs of business, industry, and the consumer. It was believed that this could be accomplished by placing the controls at the "end-of-the-pipe," be it at the tailpipe of the automobile, the tip a smokestack, or the sewer outflow pipe from a business, industry, or municipal government. Notable exceptions were in the areas of chemicals and toxic materials (see box 1.1, RCRA and TSCA), which require product and materials testing and safety certification, not simply end-of-the-pipe management.

In many respects, the implementation philosophy of the first epoch was effective in developing a strong federal implementing capacity in the form of the U.S. EPA, which in turn carried out the multiple environmental policies nationally. By doing so it helped to foster a similar capacity at the state and local level. In other respects, however, the EPA was too successful, becoming too big, cumbersome administratively, and ever-present. This contributed to the backlash against the agency and the rise of a counter-philosophy and approach that became the hallmark of second environmental epoch. The backlash gained momentum throughout the 1980s, especially during the Reagan administration in Washington, D.C., where legislative, administrative, and judicial assaults were launched by business and industry, property rights, and antienvironmental groups. While not successful enough to derail the EPA and undo the environmental movement, this did succeed in changing ideas about how best to accomplish environmental objectives (Vig and Kraft 1984; Portney 1984).

By the mid-1980s, the dominant thinking among members of Congress, business, and the broader public was that a more decentralized and collaborative approach to rule-making and goal setting—within an integrated environmental framework, where costs were accorded greater consideration in pursuing environmental gains—was a better way to accomplish the nation's ambitions environmental agenda. This shift in policy orientation was based on the assumption that the debate was no longer chiefly over the appropriateness of having environmental safeguards. What was most needed was the right opportunity and incentives for business and industry to marshal their creativity and technological know-how to meet the needs of a less-polluting and more energy efficient society. This new philosophy would balance environmental goals with private-sector costs, be more flexible in application, and be driven by "incentives" rather than governmental prescription and policing.

Economists had championed this approach for decades (Kneese and Schultze 1975; Tietenberg 1998; Meiners and Yandle 1993). But not until the second epoch did the philosophy move center stage (John 1994; National Academy of Public Administration 1994). It was tested in several pilot programs in using market mechanisms, involving stakeholders in setting rules and regulations, and working more to cajole than compel compliance with environmental mandates (Fuchs 1995; Weber 1998). This change in philosophy reached a peak in the early 1990s with the EPA's top administrators (first William Reilly, then Carol Browner) calling for a move away from pollutant-by-pollutant rules and regulations that had built up over two decades, to integrated ("multimedia"), more decentralized, and collaborative thinking and decision making.

Even before this change in philosophy had been absorbed into the culture and practices of the EPA, its state counterparts, and business and industry, an even more ambitious philosophy emerged as the environmental movement moved into its third epoch. It was the call for "sustainability," based on the conviction that more enduring solutions to the problems of environmental pollution and resource degradation were needed.

The sustainability approach envisions a complex web of human and natural systems interactions and linkages, without starting or end point. This goes well beyond the much more constricted and artificial focus characteristic of earlier thinking and policy formulation that treated air, water, land, and other pollutants separately. Linking sustainability concepts to concepts of community has particular advantages, since communities represent the social and physical expression of interdependence.

The implications for policy and action, and for social relations within a community, are profound. While it is not clear even at present how best to think about the environmental "problem" as one of sustainability, a number of efforts are under way to scope the necessary boundaries and strategies for action (Trzyna 1995; Hempel 1996). Furthermore, the absence of precision and clarity has not prevented leaders in the environmental and business communities, as well as many public officials, from embracing the cause, as evidenced by President Clinton's Council on Sustainable Development (1996).

The ultimate test of sustainability will not be in its rhetoric but in real-world applications. Important experiments are under way in a number of communities, regions, even some states across the nation, from places as small as Chattanooga, Tennessee, and South Baltimore, Maryland, to the middle-sized cities of Seattle and Pittsburgh, to entire watershed basins, such as in the Great Lakes region. Sustainability is not yet a science or a professional field, with agreed upon definitions, norms, and practices. Rather, at present it is more of a philosophy and general approach that embraces ecological values, where human health and the environment are seen as complements, where "best practices" serve as guides not formulae, and where the litmus test is the overall health of a community or ecosystem as measured by a variety of environmental and human health characteristics (Wackernagel and Rees 1996). As such, the implementation "philosophy" identifies multiple points of intervention, as well as micro and macro policy tools and guideposts. Which ones will prove the most appropriate will become clear only as the third epoch unfolds.

Information and Data Management Needs

When environmentalism emerged onto the national scene in the early 1970s, there was little question about the nature of the problem. One did not need to be an expert to appreciate that many of the nation's waterways were polluted to the point of killing off fish, and were no longer suitable for drinking and swimming. In extreme cases, their oily surface could even catch fire! Industrial, urban, and agricultural runoff was polluting many underground acquifers, rendering them useless as sources of potable water. Urban smog was an eyesore—literally—in most major urban areas, especially in the hot and dry Southwest, where Los Angeles epitomized the problem. Highly radioactive waste was accumulating at the nation's nuclear power plants with nowhere to dispose of it permanently—adding to the public's fear of possible nuclear accidents.

As a response, the passage of new pollution laws was just a first step. The EPA needed to develop, often from scratch, detailed rules and regulations for industry to follow to bring pollution within acceptable levels of public health and safety. Yet the extent of information about pollution, as well as the data gathering and measurement capabilities of the nation were quite rudimentary. While the intent of Congress to clean up the environment was clear, the practical and technical demands of gauging levels of pollution, their acute and chronic health effects, and tolerable levels of exposure were not.

Congress mandated that the regulations were to be health based, at a time when the science of epidemiology was inconclusive, dose-response rates unknown, and the understanding of the human health effects of differing amounts and duration of exposure to pollutants was modest at best. In short, a great deal of scientific and technical information was needed, and needed fast (Portney 1990a; Chivian et al. 1993). It would not be until the mid-1970s, for example, that the EPA, working with lead states such as California, was able to establish reliable monitoring methods for gauging air emissions for a community. It was also unknown which industries were generating what kinds of air and water pollution, and to what extent each could be reduced through new technology and better management, and at what cost.

Who should be responsible for gathering the needed information, and how it should be linked to regional, state, and national strategies for emissions control were hotly contested during the first epoch. In the area of water pollution, where waste was typically dumped into sewers and municipal drainage systems, an entire system of permits and emissions monitoring was developed. Even after thirty years, however, there remains only limited monitoring data on the health of the nation's rivers, streams, lakes, ponds, and reservoirs (Knopman and Smith 1993; U.S. EPA 1998b).

Air quality measures evolved with greater success, although reliable monitoring and data collection systems would not be available until late in the 1970s (Portney 1990b; U.S. EPA 1998a). Still, year-to-year variations reflect not only gains in pollution control but changing economic activity and weather patterns. Also, it is instructive and humbling to realize that even after the federal government spent over $500 million on its massive National Acid Precipitation Assessment Program in the 1980s, uncertainty remains over the mechanisms by which air emissions contribute to acid precipitation and its effect on soil chemistry and water quality.

These and similar limitations in monitoring public health and environmental quality persist and constitute one of the major barriers to monitoring environmental change and evaluating the impact of public policy efforts (Ringquist 1995; Knaap and Kim 1998). Part of the problem is that the federal government has been unable to consolidate and integrate the diverse array of environmental data programs it does have. Illustrative of this, the EPA reported in 1992 that it had identified eighty-three different environmental data programs in twenty-five separate federal agencies. Improving the nation's ability to assess interrelationships among different environmental stressors and the impact on health and the environment, and to relate this to public policy actions remains a challenge.

One of the hallmarks of the second epoch is the frequency and severity of criticism directed at environmental policy, although fewer questions are raised about the progress made to date in cleaning up the environment. Rather, critics tend to emphasize the intrusiveness of environmental policies, the substantial costs they have imposed on the private sector and on state and local governments, the inefficiency of command-and-control regulations, and the rapid rise of federal, state, and local bureaucracies associated with the nation's environmental programs (Greve and Smith 1992; Landy, Roberts, and Thomas 1994; Freeman 1997). This resulted in numerous demands that the administrative and compliance costs of regulations be weighed against pollution reduction and health gains resulting from any given rule or regulation.

The concern of the third epoch goes well beyond prescribing regulations for cleaning up pollution or conventional cost-benefit analysis of their effects. What is being asked for is a method of gauging the multiple ramifications of an action—rule, regulation, activity—within a large and complex array of possible effects, in the near term and far into the future. The level of scientific and technical data, understanding of ecological processes, and analytical capability needed for this kind of assessment is greater than ever before. Only through computer-assisted analysis and simulation will it be possible to conduct the needed analysis in most instances. With only limited data of the sort needed now available, and those with the requisite analytical skills still few in number, a great deal of new kinds of data and analytical capacity is called for, from energy and materials throughput analysis and metropolitan area footprint analysis to green accounting in business and industry.

Predominant Political/Institutional Context

The environmental policy revolution of the 1970s was guided by a set of assumptions about the capacity of state and local governments to identify and act on environmental problems, the willingness of business and industry to minimize costly remedial actions, and the capabilities of a centralized federal government to bring about substantial change in a short period of time. It was assumed, for instance, that business and industry would not voluntarily cooperate in cleanup and, indeed, that they would resist in all manner possible. Policy was designed in this adversarial context. Not surprisingly, both sides lobbied heavily as legislative proposals were formulated and debated in Congress and other venues. They also sought to shape the decisions made by administrative agencies, and they resorted often to the courts to resolve their differences and clarify rights and responsibilities (Melnick 1983; O'Leary 1993).

The second epoch reflected a desire to find a middle ground through developing new forms of collaboration and participatory policy making and rule making. With this shift came the emergence of alternative dispute resolution, extensive collaboration, negotiated rule making within the EPA, and similar processes as ways to move beyond gridlock politics and costly legal proceedings. The assumption was that bringing the key stakeholders in a policy arena together would foster greater understanding and cooperation, and allow all parties to focus on areas of shared interest and policy agreement. If nothing else, these changes launched an extraordinary era of searching for common ground and consensual solutions to deeply divisive environmental problems (Krimsky and Plough 1988; National Research Council 1989; Mazmanian and Morell 1994; Kraft 1994; Weber 1998; Williams and Matheny 1995).

As important as these new decision-making processes are as educational devices and ways of resolving disputes, they have not fully eliminated the suspicions and conflicts among contending parties, nor are they always well suited to resolving fundamental conflicts. Indeed, the confrontational politics used during the Republican-controlled 104th and 105th Congresses (1995–1998) to weaken environmental policy, and the strong countermovement by a rejuvenated environmental community, attest to the deeper cultural persistence of politics as usual (Bosso 1999; Kraft 1999). It would appear that the willingness by the parties to environmental conflicts to use alternative mechanisms for resolving disputes in the second epoch reflects only a partial and strategic commitment on their parts, rather than a fundamental rejection of the older adversarial form of political decision making.

For the third epoch, now in its formative stage, collaboration and cooperation among all affected stakeholders and incentive-based methods of policy implementation are promoted as the preferred approaches for both philosophical and instrumental reasons (Maser 1997). What began as experimentation in epoch two is being embraced more deeply in order to reach and assure genuine community-based sustainability. In this vein the President's Council on Sustainable Development focused on this dimension in its 1996 report to the nation. With remarkable unanimity, the council's members called for a new generation of flexible, consensual environmental policies that would maximize economic welfare while achieving more effective and efficient environmental protection. The United States, council members said, "must change by moving from conflict to collaboration and adopting stewardship and individual responsibility as tenets by which to live" (PCSD 1996, 1). Such a statement may reflect more idealism than realism, but it does indicate the kind of political values that are beginning to infuse third-epoch thinking and the effort to seek new forms of governance, starting at the local community level (Hempel 1998).

Examples of this new approach can already be found (Sexton, et al. 1999; Sitarz 1998, ch. 10). For several years in the late 1990s, the EPA had a formal pilot program in operation in South Baltimore, Maryland, to form a Community Environmental Partnership. "Community-based environmental protection," as the EPA calls it, brings together community residents, businesses, and government representatives to identify local environmental problems and to develop a consensus action plan and priorities or "ways to foster health, environment, and sustainable economic development in the community" (U.S. EPA 1996a and 1997). These efforts do not eliminate applicable regulatory actions for the parties involved. Rather they encourage local pollution prevention and environmental restoration efforts and citizen involvement in decision making.

The EPA also provides technical training and scientific expertise to communities through a new Office of Sustainable Ecosystems and Communities (OSEC), created in 1994. OSEC has been developing training programs in mediation and negotiation for EPA regional staff, assisting communities in developing local cultural profiles in the context of conservation assessments and planning, and compiling indicators for community sustainability. The office also has established a Sustainable Ecosystems and Communities Clearinghouse, which provides information through the EPA's World Wide Web site, as well as through newsletters and fax-on-demand services. The Web site provides access to information about ecosystem science, the economics of sustainability, community involvement, and tools and guidance for integration of such knowledge, including indicators of sustainability, case studies, and handbooks for citizens.[4] OSEC defines its role in this new approach that is representative of the third epoch as one of broker, educator, advocate, analyst, and financier to "foster the implementation of an integrated, geographic approach to environmental protection that emphasizes ecological integrity and the associated improvements in economic sustainability and quality of life" (U.S. EPA 1996b).

Analysts examining these trends hail the potential of public education campaigns, collaborative decision making, public-private partnerships, and continued and increasing reliance on market incentives (John 1994; National Academy of Public Administration 1995; Press and Mazmanian 1997). A State of the Environment report for the international Commission for Environmental Cooperation (CEC) similarly highlights the attractiveness of nonregulatory approaches of this kind and documents their use in Canada, Mexico, and the United States over the past several years (CEC 1999).

The transition is far from complete, however, and for the present we have an unusual hybrid form of environmental regulation in which new approaches are placed alongside or overlying the old. While the potential for significant improvement has been demonstrated when the right conditions present themselves, the persistence of the formal apparatus of command-and-control regulation in the major environmental protection statutes and their attendant regulations remains, and continues to be appropriate in many instances.

What Follows

The chapters that follow provide a combination of further clarification of the conceptual issues underlying the changes in the environmental movement and concrete evidence of the two major transformations under way: from the first to second as well as second to third epochs. The dates dividing the epochs are never as clear and crisp in practice as described analytically and laid out in table 1.1, and the timing and overlapping nature of the transformations fit some cases, such as air and water, better than others, such as land use and transportation. Nonetheless, the broad outlines will be revealed and the influence of the growth and learning across the wide array of environmental arenas—trial and error in one epoch, new ideas emerging and becoming dominant in the next, with revised goals and proposed new strategies—are striking. Understanding the environmental movement as it evolves through the three epochs is valuable in understanding the key dimension of societal-level change in general and the transformation of environmental policy in particular. This knowledge should find practical application among those seeking to foster and accelerate the drive toward sustainability.

Meanwhile, as scholars and researchers, we are only beginning our task. To the extent that the epochs framework proves useful in guiding analysis and helping to provide understanding of the changes past and present, and insights into those we can look for in the future through the initial set of cases the book presents, the framework will beg even more questions of verification and possible applications. The concluding chapter will address these issues.

Chapters 3 through 8 provide an introduction and overview of a cross section of major environmental policy areas in light of the epochs framework laid out above. This analysis begins with cases that illuminate the transition from the first to second epoch followed by cases of transitions under way into the third epoch. The cases were chosen because they highlighted (and provided a test of) dimensions of the transition framework combined with their substantive importance as an environmental concern. Consequently, the further one moves through the book, the greater is the shift in emphasis on just environmental protection to sustainability of entire communities and regions, within which environmental concerns become embedded in a far more comprehensive framework and discussion, that is, sustainability.

With this organization in mind, and before launching into the particular cases, the newness, complexity, and ambiguities surrounding the notion of sustainable communities necessitates further discussion and amplification. Therefore chapter 2, by Lamont Hempel, is devoted to providing a broad overview of the origins and implications of the emergent focus on sustainability and why it is having the effect today of repositioning and galvanizing the environmental movement. The chapter serves to underscore the significant departure from the first and second environmental epochs that sustainability represents and thus the transformational challenge it poses for the environmental movement.

Section two of the book provides examples of important community- and regionally-based efforts at addressing conventional environmental pollution problems. Chapter 3, by Daniel Mazmanian, examines clean air regulation, using the case of the transformation in conceptual approach and practice in Southern California, the most heavily polluted urban region in the nation. In chapter 4, Michael Kraft and Bruce Johnson describe extended and committed efforts to use collaborative decision making at the local level to address both point and nonpoint source pollution management under the Clean Water Act. They illuminate the promise and pitfalls of the new approach through an examination of regional watershed management in the Fox-Wolf River Basin in Northeastern Wisconsin. The case illustrates the inherent constraints of conventional environmental regulation and the attractiveness—and limitations—of incentive-based systems and voluntary agreements. Chapter 5 shifts the focus to land-use policy, which historically has been fashioned in terms of parks, recreation, preservation, and aesthetics, but only indirectly at best, sustainability. Here Daniel Press helps us understand the historical context of land-use policy, with emphasis on the important role of local values, interests, and administrative mechanisms (what political scientists call institutions), and how these have evolved through the epochs, albeit on a very different timescale than painted in chapter 1. Press focuses on open space preservation in California, and he seeks to learn why some communities succeed in preserving open space while others do not.

The third section of the book draws on new and encompassing efforts, where multimedia, multisector, and multistakeholder strategies are being developed within communities and regions based on a comprehensive view and long-term sustainability. The three chapters within this section examine very different contexts within which sustainability is being adopted and innovative practices developed. Chapter 6 begins with the urban, rust-belt context of the city of Pittsburgh. Franklin Tugwell, Andrew McElwaine, and Michele Fetting trace the history of Pittsburgh's efforts to deal with a host of environmental problems, focusing on recent efforts to promote a greener city through use of public-private partnerships, and the role played by "change agents" or policy entrepreneurs. They set forth the achievements of these remarkable activities, yet they also highlight the continuing obstacles (in Pittsburgh and elsewhere) to moving more resolutely into the epoch of sustainability. Shifting from a single urban setting to the level of national infrastructure, in chapter 7 Thomas Horan, Hank Dittmar, and Daniel Jordan set out the technical, economic, and political foundations for a sustainable U.S. transportation policy. Building on the achievements of the innovative 1991 Intermodal Surface Transportation Efficiency Act (ISTEA), they examine the potential and limitations for a holistic transportation policy consistent with the idea of sustainable communities. Concluding this section, chapter 8, by Barry Rabe, assesses one of the most ambitious efforts to date to move environmental policy through the three epochs. This is the case of the long-term cleanup and restoration efforts within the Great Lakes Basin.

In the concluding chapter 9, by Kraft and Mazmanian, the discussion returns to the opening issues of the theory-practice nexus in assessing the extent to which the epochs framework helps us understand the profound changes in the environmental movement past, present, and as we look into the future. Special attention is given to efforts to clarify the concept of sustainability at the community level in its environmental, social, and political dimensions. The chapter addresses the political and participatory dimensions, indeed requisites, of the transformation to sustainability implicit throughout all the illustrative cases.

All of the chapters incorporate an interdisciplinary orientation appropriate to the subject matter. They focus on important illustrations that provide evidence of what works and what does not at the local and regional level, and why. These will not be final words on the subjects addressed. But they do move the discussion forward and help identify the conditions for successful environmental policy development and implementation. Thus they help to define the basis for policy prescriptions for localities and regions seeking to initiate strategies for building truly enduring sustainable communities.

Notes

1. A recent compilation by Jonathan Collett and Stephen Karakashian, eds., *Greening the College Curriculum* (Washington, D.C.: Island Press, 1996), offers extensive annotated bibliographies of scholarship in anthropology, biology, economics, geography, history, literature, media and journalism, philosophy, political science, and religion. In addition, see Robert Paehlke's comprehensive *Conservation and Environmentalism: An Encyclopedia* (New York: Garland Publishing, 1995).

2. Likewise, a population that distributes its resources equitably can avoid the extremes of high per capita consumption by the rich and borderline survival (in terms of per capita consumption) by the poor.

3. The life of the EPA was to be neither steady or smooth. The pattern of rapid growth and significant institutional learning and capacity building in 1970s was curtailed in the early 1980s as criticism of the agency mounted and a shift in the political climate led to greater demands for balancing environmental protection and economic impacts. Sharp cuts in its budget and staff resulted, causing considerable administrative turbulence and erosion of staff capabilities and morale (Vig and Kraft 1984; Harris and Milkis 1996). For an overview of the changes in the EPA's budget and staff from the 1970s to the late 1990s, see Vig and Kraft (1999).

4. The Web site for OSEC is: www.epa.gov/ecocommunity.

References

Axelrod, Regina S., and Norman J. Vig. 1999. "The European Union as an Environmental Governance System." In *The Global Environment,* ed. Vig and Axelrod. Pp. 72–97.

Bartone, Carl, Janis Berstein, Josef Leitmann, and Jochen Eigen. 1994. "Toward Environmental Strategies for Cities: Policy Considerations for Urban Environmental Management in Developing Countries." Washington, DC: The World Bank.

Bosso, Christopher J. 1999. "Facing the Future: Environmentalists and the New Political Landscape." In *Environmental Policy,* eds. Vig and Kraft, Pp. 55–76.

Bryner, Gary C. 1995. *Blue Skies, Green Politics: The Clean Air Act of 1990 and Its Implementation,* 2nd ed. Washington, DC: CQ Press.

Caldwell, Lynton K. 1982. *Science and the National Environmental Policy Act: Redirecting Policy Through Procedural Reform.* University, AL: University of Alabama Press.

Chivian, Eric, Michael McCally, Howard Hu, and Andrew Haines. 1993. *Critical Condition: Human Health and the Environment.* Cambridge, MA: MIT Press.

Cobb, Clifford, Ted Halstead, and Jonathan Rowe. 1995. "If the GDP Is Up, Why Is America Down?" *The Atlantic Monthly* (October):59–78.

Commission for Environmental Cooperation. 1999. *Taking Stock: Sustainability and the State of the North American Environment.* Montreal: CEC.

Cortner, Hanna J, and Margaret A. Moote. 1998. *The Politics of Ecosystem Management.* Washington, DC: Island Press.

Council on Environmental Quality. 1997. *Environmental Quality: 25th Anniversary Report of the Council on Environmental Quality.* Washington, DC: Government Printing Office.

Daly, Herman E., and John B. Cobb, Jr. 1989. *For the Common Good: Redirecting the Economy Toward Community, the Environment, and a Sustainable Future.* Boston: Beacon Press.

Daly, Herman E. 1977. *Steady-State Economics.* New York: W.H. Freeman.

Davies, J. Clarence, ed. 1996. *Comparing Environmental Risks: Tools for Setting Government Priorities.* Washington, DC: Resources for the Future.

Davies, J. Clarence, and Jan Mazurek. 1998. *Pollution Control in the United States: Evaluating the System.* Washington, DC: Resources for the Future.

Dunlap, Riley E. 1995. "Public Opinion and Environmental Policy." In *Environmental Politics and Policy: Theories and Evidence.* 2nd ed., ed. James P. Lester. Durham, NC: Duke University Press. Pp. 63–114.

Fiorino, Daniel. 1996. "Toward a New System of Environmental Regulation: The Case for an Industry Sector Approach." *Environmental Law,* Vol. 26 (Summer): 457–488.

Freeman, A. Myrick III. 1997. "Economics, Incentives, and Environmental Regulation." In *Environmental Policy in the 1990s*, eds. Vig and Kraft. Pp. 187–207.

Fuchs, Doris Andrea. 1995. "Incentive Based Approaches in Environmental Policy—How Little We Know." Claremont, CA: Center for Politics and Economics (March).

Geis, Don, and Tammy Kutzmark. 1995. "Developing Sustainable Communities: The Future Is Now." *Public Management* (August):4–13.

Gray, Rob (with Jan Bebbington and Diane Walters). 1993. *Accounting for the Environment*. Princeton, NJ: Markus Wiener Publishers.

Greve, Michael S., and Fred L. Smith, Jr., eds. 1992. *Environmental Politics: Public Costs, Private Rewards*. New York: Praeger.

Hamilton, Michael S., ed. 1990. *Regulatory Federalism, Natural Resources, and Environmental Management*. Washington, DC: American Society for Public Administration.

Hempel, C. Lamont. 1996. *Environmental Governance: The Global Challenge*. Washington, DC: Island Press.

Hempel, C. Lamont. 1998. "Sustainable Communities: From Vision to Action." Claremont, CA: The Claremont Graduate University.

Hren, Benedict, Nick Bartolomeo, and Michael Signer. 1995. "Securing Your Future: Pathways to Community Sustainability." Gaithersburg, MD: The Izaak Walton League of America.

Janicke, Martin, Harald Monch, and Manfred Binder. 1993. "Ecological Aspects of Structural Change." Berlin: Free University of Berlin, FFU rep 93–1.

John, DeWitt. 1994. *Civic Environmentalism: Alternatives to Regulation in States and Communities*. Washington, DC: CQ Press.

Knaap, Gerrit J., and Tschangho John Kim, eds. 1998. *Environmental Program Evaluation: A Primer*. Champaign, IL: University of Illinois Press.

Kneese, Allen,and Charles Schultze. 1975. *Pollution, Prices and Public Policy*. Washington, DC: Brookings Institution.

Knopman, Debra S., and Richard A. Smith. 1993. "Twenty Years of the Clean Water Act." *Environment* 35 (January-February): 17–20, 34–41.

Kraft, Machael E. 1999. "Environmental Policy in Congress: From Consensus to Gridlock" In *Environmental Policy*. ed. Vig and Kraft. Pp. 121–144.

Kraft, Michael E. 1996. *Environmental Policy and Politics: Toward the Twenty-First Century*. New York: HarperCollins.

Kraft, Michael E. 1994. "Searching for Policy Success: Reinventing the Politics of Site Remediation." *The Environmental Professional* 16 (September): 245–253.

Krimsky, Sheldon, and Alonzo Plough. 1988. *Environmental Hazards: Communicating Risks as a Social Process*. Dover, MA: Auburn House.

Landy, Marc, Marc Roberts, and Stephen Thomas. 1994. *The Environmental Protection Agency: Asking the Wrong Questions,* expanded ed. New York: Oxford University Press.

Lester, James P. 1995. "Federalism and State Environmental Policy." In *Environmental Politics and Policy: Theories and Evidence,* 2nd ed., ed. James P. Lester. Durham, NC: Duke University Press. Pp. 39–60.

Liefferink, Duncan. 1999. "The Dutch National Plan for Sustainable Society." In *The Global Environment,* ed. Vig and Axelrod. Pp. 256–278.

Lowry, William. 1992. *The Dimensions of Federalism: State Governments and Pollution Control Policy.* Durham, NC: Duke University Press.

Maser, Chris, 1997. *Sustainable Community Development: Principles and Concepts.* Delray Beach, Florida: St. Lucie Press.

Mazmanian, Daniel, and David Morell. 1994. "The 'NIMBY' Syndrome: Facility Siting and the Failure of Democratic Discourse." In *Environmental Policy in the 1990s: Toward a New Agenda,* 2nd ed., ed. Norman J. Vig and Michael E. Kraft. Washington, DC: CQ Press. Pp. 233–249.

Mazmanian, Daniel, and David Morell. 1992. *Beyond Superfailure: America's Toxics Policy for the 1990s.* Boulder, CO: Westview Press.

Mazmanian, Daniel, and David Morell. 1991. "EPA: Coping With the New Political Economic Order." *Environmental Law,* 21, 4:1477–1491.

Mazmanian, Daniel, and David Morell. 1988. "The Elusive Pursuit of Toxics Management." *The Public Interest* 90 (Winter): 81–98.

Mazmanian, Daniel, and Paul Sabatier. 1989. *Implementation and Public Policy,* with a new postscript. University Press of America.

Meiners, Roger, and Bruce Yandle, eds. 1993. *Taking the Environment Seriously.* Boston: Rowman & Littlefield Publishers, Inc.

Melnick, R. Shep. 1983. *Regulation and the Courts: The Case of the Clean Air Act.* Washington, DC: Brookings Institution.

Milbrath, Lester W. 1989. *Envisioning a Sustainable Society: Learning Our Way Out.* Albany: State University of New York Press.

Minard, Richard. 1996. "CRA and the States: History, Politics, Results." In *Comparing Environmental Risks,* ed. J. Clarence Davies. Pp. 23–61.

National Academy of Public Administration. 1995. *Setting Priorities, Getting Results: A New Direction for EPA.* Washington, DC: National Academy of Public Administration.

National Academy of Public Administration. 1994. *The Environment Goes to Market: The Implementation of Economic Incentives for Pollution Control.* Washington, DC: National Academy of Public Administration.

National Commission on the Environment. 1993. *Choosing a Sustainable Future: The Report of the National Commission on the Environment.* Washington, DC: Island Press.

National Research Council. 1994. *Science and Judgment in Risk Assessment.* Washington, DC: National Academy Press.

National Research Council. 1989. *Improving Risk Communication.* Washington, DC: National Academy Press.

Nijkamp, Peter, and Adriaan Perrels. 1994. *Sustainable Cities in Europe: A Comparative Analysis of Urban Energy—Environmental Policies.* London: Earthscan Publications.

O'Leary, Rosemary. 1993. *Environmental Change: Federal Courts and the EPA.* Philadelphia: Temple University Press.

Ophuls, William, and A. Stephen Boyan, Jr. 1992. *Ecology and the Politics of Scarcity Revisited.* New York: W.H. Freeman.

Portney, Paul R., ed. 1984. *Natural Resources and the Environment: The Reagan Approach.* Washington, DC: Urban Institute Press.

Portney, Paul R., ed. 1990a. *Public Policies for Environmental Protection.* Washington, DC: Resources for the Future.

Portney, Paul R. 1990b. "Air Pollution Policy." In *Public Policies for Environmental Protection,* ed. Portney. Pp. 27–96.

President's Council on Sustainable Development. 1996. *Sustainable America: A New Consensus for Prosperity, Opportunity, and a Healthy Environment for the Future.* Washington, DC: President's Council on Sustainable Development, February.

Press, Daniel, and Daniel Mazmanian. 1997. "The Greening of Industry: Achievement and Potential." In *Environmental Policy in the 1990s,* 3rd edition, ed. Vig and Kraft. Pp. 255–277.

Rabe, Barry. 1997. "Power to the States: The Promise and Pitfalls of Decentralization." In *Environmental Policy in the 1990s,* 3rd edition, ed. Vig and Kraft. Pp. 31–52.

Ringquist, Evan J. 1995. "Evaluating Environmental Policy Outcomes." In *Environmental Politics and Policy,* 2nd edition, ed. James P. Lester. Durham: Duke University Press. Pp. 303–327.

Ringquist, Evan J. 1993. *Environmental Protection at the State Level: Politics and Progress in Controlling Pollution.* Armonk, NY: M. E. Sharpe.

Rosenbaum, Walter. 1997. "The EPA at Risk: Conflicts over Institutional Reform." In *Environmental Policy in the 1990s,* ed. Vig and Kraft. Pp. 143–167.

Schmidheiny, Stephan. 1992. *Changing Course: A Global Business Perspective on Development and the Environment.* Cambridge, MA: MIT Press.

Schneider, Anne Larason., and Helen Ingram. 1997. *Policy Design for Democracy.* Lawrence, KS: University of Kansas Press.

Schneider, Anne L., and Helen Ingram. 1990. "Policy Design: Elements, Premises, and Strategies." In *Policy Theory and Policy Evaluation: Concepts, Knowledge, Causes, and Norms,* edited by Stuart S. Nagel. Westport, CT: Greenwood Press. Pp. 77–101.

Schon, Donald A. and Martin Rein. 1994. *Frame Reflection: Toward the Resolution of Intractable Policy Controversies.* New York: Basic Books.

Sexton, Ken, Alfred A. Marcus, K. William Easter, and Timothy D. Burkhardt, eds. 1999. *Better Environmental Decisions: Strategies for Governments, Businesses, and Communities.* Washington, D.C.: Island Press.

Sitarz, Daniel, ed. 1998. *Sustainable America: America's Environment, Economy and Society in the 21st Century.* Carbondale, IL: EarthPress.

Smart, Bruce. 1992. *Beyond Compliance: A New Industry View of the Environment.* Washington, DC: World Resources Institute.

Stavins, Robert. 1991. "Project 88—Round II, Incentives for Action: Designing Market-Based Environmental Strategies." Washington, DC: A public policy study sponsored by Senator Timothy E. Wirth and Senator John Heinz, May.

Tietenberg, Tom. 1998. *Environmental Economics and Policy,* 2nd edition. Reading, MA: Addison-Wesley.

Trzyna, Thaddeous, ed. 1995. *A Sustainable World: Defining and Measuring Sustainable Development.* Sacramento, CA: California Institute of Public Affairs.

United Nations. 1993. *Agenda 21: Programme of Action from Rio.* NY: United Nations.

U.S. Environmental Protection Agency. 1998a. *National Air Quality and Emissions Trends Report, 1996.* Research Triangle Park, NC: Office of Air Quality Planning and Standards, EPA 454/R-97-013, January.

U.S. Environmental Protection Agency. 1998b. *National Water Quality Inventory: 1996 Report to Congress.* Washington, DC: Office of Water, EPA 841-F-97-003, April.

U.S. Environmental Protection Agency. 1997. *EPA Strategic Plan.* Washington, DC: Office of the Chief Financial Officer, EPA 190-R-97-002, September.

U.S. Environmental Protection Agency. 1996a. "Community Partnerships for Environmental Action: A New Approach to Environmental Protection." Washington, DC: Office of Pollution Protection and Toxics, U.S. EPA.

U.S. Environmental Protection Agency. 1996b. "Information Packet." Washington, DC: Office of Sustainable Ecosystems and Communities, U.S. EPA.

U.S. Environmental Protection Agency. 1990. *Reducing Risk: Setting Priorities and Strategies for Environmental Protection.* Washington, DC: EPA, Science Advisory Board, September.

Van Dieren, Wouter, ed. 1995. *Taking Nature Into Account: Toward a Sustainable National Income.* New York: Springer-Verlag.

Vig, Norman J., and Regina S. Axelrod, eds. 1999. *The Global Environment: Institutions, Law, and Policy.* Washington, D.C.: C.Q. Press.

Vig, Norman J., and Michael E. Kraft, eds. 1999. *Environmental Policy: New Directions for the 21st Century.* Washington, DC: CQ Press.

Vig, Norman J., and Michael E. Kraft, eds. 1997. *Environmental Policy in the 1990s: Reform or Reaction?,* 3rd ed. Washington, DC: CQ Press.

Vig, Norman J., and Michael E. Kraft, eds. 1984. *Environmental Policy in the 1980s: Reagan's New Agenda*. Washington, DC: CQ Press.

Wackernagel, Mathis, and William Rees. 1996. *Our Ecological Footprint: Reducing Human Impact on the Earth*. Philadelphia, PA: New Society Publishers.

Weber, Edward P. 1998. *Pluralism by the Rules: Conflict and Cooperation in Environmental Regulation*. Washington, DC: Georgetown University Press.

Westra, Laura. 1994. *An Environmental Proposal for Ethics: The Principle of Integrity*. Lanham, MD: Rowman & Littlefield.

Williams Bruce, and Albert Matheny. 1995. *Democracy, Dialogue, and Environmental Disputes: The Contested Languages of Social Regulation*. New Haven: Yale University Press.

World Commission on Environment and Development (Brundtland Commission). 1987. *Our Common Future*. New York: Oxford University Press.

Yaffee, Steven L., Ali F. Phillips, Irene C. Frentz, Paul W. Hardy, Sussanne M. Maleki, and Barbara E. Thorpe. 1996. *Ecosystem Management in the United States: An Assessment of Current Experience*. Washington, DC: Island Press.

2
Conceptual and Analytical Challenges in Building Sustainable Communities

Lamont C. Hempel

The emerging sustainability ethic may be more interesting for what it implies about politics than for what it promises about ecology. Whether viewed as a universal truth about intergenerational justice and interspecies harmony or as a half-baked attempt to make a virtue out of mere endurance, the idea of sustainability has successfully invaded the rhetoric, if not the substance, of policy debate. What remains to be seen is whether the concept will fundamentally influence policy formation and become, along with efficiency, effectiveness, and equity, a leading criterion in the design and evaluation of policies and institutions.

Growing interest in sustainability can be explained in a variety of ways. In part, it may be a response to the "maelstrom of perpetual disintegration"(Berman 1982, 15) that postmodernists say threatens global ecology, cultural diversity, and democratic institutions. In a related manner, perhaps it is a reaction to the dizzying changes wrought by advancing technology—a plea for continuity to withstand Toffler's (1970) "future shock." Conceivably, growing interest in sustainability is a result of social learning about environmental destruction and about the foregone opportunities that rapid development and profligate consumption promote. More likely, at least for cynics, it is an outgrowth of nostalgia—however exaggerated—for simpler times, less traffic congestion, more open space, and an abiding sense of place.

This chapter examines the implications of sustainability concepts and their application in real-world communities. It explores the construction of meaning in beliefs about sustainability, and the conceptual promise and analytical difficulties that those beliefs create for local and regional policy making. Defining the term *sustainability* with precision is less important for this purpose than understanding the characteristics and goals of the social movement it has fostered. Like other transformative ideas, the concept of sustainability promises to remake the world through reflection and choice, but its potential to engage people's hopes, imagination, and sense of responsibility may depend more on strategic uses of ambiguity than on conceptual precision and clarity. Mobilizing ideas appears to be most effective when they serve as condensational symbols that defy narrow definition, encourage coalition building among diverse interests, and permit just enough comprehension and social absorption to promote convergent political acts. The symbol of sustainability, arguably, is sufficiently ambiguous to be embraced by diverse interests, yet coherent enough to inspire movement in a particular direction.

Although the evolution of environmental policy has been pointing in the direction of sustainability concepts for many years, the path has seldom been straight and narrow. As shown in chapter 1, the historical development spans three overlapping epochs, with the latest one yet to emerge fully. Each epoch represents a different environmental management philosophy—from national command-and-control strategies to increased reliance on state and local management and market-based approaches and, eventually, to community design based on sustainability. Succession has been driven by a combination of policy failures, political leadership changes, and improvements in our understanding of how social, economic, and environmental goals interact. It took a series of painful lessons to convince many in the environmental movement that cleaning up the air, water, and soil, protecting other species, and conserving natural resources would require more attention to the needs of communities and less reliance on national legislation and single-media pollution strategies. An important lesson about sustainability is that lasting gains in quality of life cannot be achieved without effective integration of environmental, social, and economic goals at the community and regional level.

Conceptual Challenges

No matter what object of sustainability is measured, there is a range of time across which sustainability is not achievable. Communities that may be sustainable over the time frames of modern history—that is, centuries—become faint punctuation marks in the eons of geological history. Even a sustainable Earth eventually succumbs to entropy, asteroid collisions, or other astronomical cataclysms. Nothing is permanent in a physical sense, including our solar system. Accordingly, human communities cannot be sustainable in any strict sense of the term.

Practically speaking, of course, the concept of sustainable community is temporally acceptable for most human endeavors. In the short run, definitions of sustainability are problematic only insofar as they move from theory to practice. Within the relatively short time frames in which community decision makers operate, the goals of sustainability cease to appear paradoxical. The real difficulty lies not in the relativity of the concept but in its operationalization. Applying sustainability criteria to everyday matters of public policy, business management, and personal consumption is fraught with conceptual and moral hazards. Equally challenging is the problem of envisioning a living sustainable community in the absence of any concrete examples. And even if persuasive examples could be found and unifying visions embraced, few contemporary institutions or systems of governance appear flexible enough to carry out such visions in a timely manner. Building communities in which environmental quality, social justice, and economic vitality cohere in some sustained fashion requires a rare combination of long-range foresight and short-term adaptability.

The Process of Sustainability
For hard-bitten realists, the goal of community sustainability is chimerical. Most definitions of *sustainability* and of *community* yield enough ambiguity to frustrate even the most careful attempts at conceptual linkage and integration.[1] We can reduce some of that ambiguity by limiting the use of the term *community* to geographic associations of people who share some social, political, historical, and economic interests. But *sustainability* is much harder to qualify in this way. It is a relative, context-intensive term, in the sense that its meaning depends more on dynamic understandings of process than on fixed notions of semantics and final interpretation. As Folke and Kaberger (1991, 289) observe, "It is not meaningful to measure the absolute sustainability of a society at any point in time." We will have to settle for relative assessments of sustainability. Sustainability as permanence is not an op-

tion. Given the second law of thermodynamics, the very idea of sustainability can be called into question. And yet there is something deeply satisfying about the concept, at least in the minds of many environmentalists, policy makers, business leaders, and public intellectuals.[2]

Efforts to define and frame concepts of sustainability have typically stressed the principles of support, persistence, balance, and, most important, resilience. A sample of these and other common approaches and definitions is presented in box 2.1. Although some define sustainability in terms of stability, ecologists note that ecosystems with low stability can still exhibit high resilience, thereby suggesting that sustainability has more to do with endurance than with order.[3] Critics, however, respond that if the essence of sustainability is endurance, the concept is neither intellectually interesting nor ethically supportable. Paul Treanor (1996, 6), for example, asserts that "no moral argument can justify the continued existence of existing." Because sustainability assigns value to duration, it provides a basis, in Treanor's view, for radical traditionalism and inflexible forms of political conservatism.

Sustainable Development

By far the most common usage of the word *sustainable* is in combination with the word *development*. Formally introduced in the *World Conservation Strategy* (IUCN 1980), popularized by the Brundtland Commission (World Commission on Environment and Development 1987), and diplomatically embraced at the Earth Summit in 1992, sustainable development has become the "guide star" of international efforts to reconcile economic and ecological imperatives.

Despite its popularity within the U.N. system and across a wide range of opinion leaders, the concept of sustainable development has attracted considerable opposition from environmental groups, most of it centering on the word *development*.[4] William Ophuls (1996, 34) is typical of those who criticize the linkage to development:

Sustainable development is an oxymoron. Modern political economy in any form is unsustainable, precisely because it involves "development"—that is, more and more people consuming more and more goods with the aid of ever more powerful technologies . . . such an economy is based on stolen goods, deferred payments,

Box 2.1
Sustainability: A sampling of definitions and conceptual approaches

---

• Sustenance—that which supports life.
To support without collapse; to keep up, carry, or withstand (Webster's *Dictionary*).
• A nondeclining utility function or nondeclining capital; nondeclining mean human welfare over time (Pearce, Markandya, and Barbier 1989).
• Resilience—ability to maintain structural integrity, form, and patterns of behavior in the midst of disturbance (Common 1995).
• Securing people's well-being within the means of Nature (Wackernagel 1997)
• A process of creation, maintenance, and renewal that persists in balance with the process of decline, death, and decay (Hempel 1992).
• Sustainability is about opportunity and the preservation of meaningful choice—"leaving future generations as many or more opportunities as we have" (Serageldin 1995).
• A condition in which social systems and natural systems thrive together indefinitely (Euston 1995).
• [Sustainable Development] ". . . the complex of activities that can be expected to improve the human condition in such a manner that the improvement can be maintained" (Munro 1995).
• [Sustainable Development] ". . . development that meets the needs of the present without compromising the ability of future generations to meet their own needs" (Brundtland Commission 1987).
• [Sustainable Development] ". . . the system does not cause harm to other systems, both in space and time; the system maintains living standards at a level that does not cause physical discomfort or social discontent to the human component; within the system life-support ecological components are maintained at levels of current conditions, or better" (Voinov and Smith 1998).
• [Sustainable Communities] "healthy communities where natural and historic resources are preserved, jobs are available, sprawl is contained, neighborhoods are secure, education is lifelong, transportation and health care are accessible, and all citizens have opportunities to improve the quality of their lives" (President's Council on Sustainable Development 1996, 12).

and hidden costs; it continues to exist or even thrive today only because we do not account for what we steal from nature or for what posterity will have to pay for our pleasures or for what we sweep under the ecological carpet.

Others have emphasized the ways in which debt, trade, and foreign aid have altered the meaning of development and given it a peculiarly Western, industrial focus that fosters accelerated destruction of biological and cultural diversity, particularly in "underdeveloped" countries and regions.[5]

Focus on Community

Partly in response to dissatisfaction with the sustainable development concept and partly in response to growing concerns about urban quality of life, a splinter movement of sorts has arisen in an effort to focus sustainability strategies on the social, economic, and ecological well-being of communities. Participants in this movement define community sustainability in ways that highlight relationships between local quality of life and local or regional levels of population, consumption, political participation, and commitment to intertemporal equity. A sustainable community is one in which economic vitality, ecological integrity, civic democracy, and social well-being are linked in complementary fashion, thereby fostering a high quality of life and a strong sense of reciprocal obligation among its members. Ideally, sustainable communities have levels of pollution, consumption, and population size that are in keeping with regional carrying capacity; their members share an ethic of responsibility to one another and to future generations; prices of their goods and services reflect, where practical, the full social costs of their provision; equity mitigation measures protect the poor from the regressive impacts of full-cost pricing; community systems of education, governance, and civic leadership encourage informed democratic deliberation; and the design of markets, transport, land use, and architecture enhances neighborhood livability and preserves ecological integrity (Hempel 1996b and 1998).

The sustainable communities movement tends to be highly decentralized and inclusive, drawing together a growing number of environmentalists, urban public health groups, growth management advocates, civic leaders, municipal planning and redevelopment agencies, environmental justice activists, proponents of local control in government, neighborhood associations, and scholars interested in interdisciplinary approaches to urban problem solving. Although becoming visible in dozens of countries, the movement has not yet achieved any formal coalescence or unification in the organized pursuit of particular policies or institutional reforms. The dispersed nature of the movement is perhaps best illustrated by the more than 400 sustainability Web sites that are now linked electronically on the Internet.[6]

Evolving Frameworks

The historical roots of the sustainable communities movement in North America can be traced to many different sources, some major and some relatively minor in importance. A partial list of formative developments in the evolution of this movement is provided in box 2.2.

Few individuals have done more to lay the groundwork for today's sustainable community movement than Patrick Geddes and Lewis Mumford. Geddes's *Cities in Evolution* (1915) stressed the integration of environmental protection and social organization in urban design. Noting the corrosive effects of industrialization on geographic community, Geddes observed that "[s]uch swift multiplication of the quantity of life, with corresponding swift exhaustion of the material resources on which life depends has been too much . . . "(1915, 52). His plea for regional planning to combat the "waste and dissipation" of Nature and community strongly influenced the young Lewis Mumford (Luccarelli 1995, 16–28), who went on to become a leader in the 1920s Regional Planning Association of America. Mumford's *Sticks and Stones* (1924) and *The Golden Day* (1926) were among the earliest attempts to address jointly the social and environmental consequences of the loss of community in a "machine civilization." Portraying colonial New England as a promising model for the contemporary design of human settlements, Mumford attempted to reconstruct the communitarian social tradition in a way that would be appropriate for the twentieth century. According to Luccarelli (1995, 49–50), Mumford believed that

the attainment of a genuine sense of place, grounded in relation to nature, parallels—and encourages—community. In the early New England that Mumford

Box 2.2
Sustainable communities: An issue evolution framework

• Garden City movement, led by Ebenezer Howard (1898) at the dawn of the twentieth century.
• Bioregional planning and design insights of Patrick Geddes (1915), Ian McHarg (1969), Kirkpatrick Sale (1985), and the Regional Planning Association of America (Sussman 1976).
• American New Towns movement (e.g., Reston, VA; Radburn, NJ; Columbia, MD) that began in the 1920s.
• Grassroots communitarian movements of the late 1940s (e.g., Fellowship of Intentional Communities, established in 1948).
• Great Society urban programs of Lyndon Johnson in the 1960s.
• The decline of faith in technological progress as a solution to urban problems (e.g., Mumford 1964).
• Spaceship earth idea (Fuller 1969).
• "Limits to growth" arguments of the late 1960s and early 1970s (e.g., Meadows et al. 1972).
• Studies of the resilience of ecological communities (Holling 1973).
• Local self-reliance and appropriate technology movements of the 1970s and early 1980s (e.g., Morris 1982).
• Urban Ecology and Eco-City movement (e.g., Register 1987, Engwicht 1993).
• The strategic coupling of environment and development interests in the sustainable development dialogue of the 1980s (e.g., the Brundtland Commission, 1987) and of the 1990s (e.g., the Earth Summit, 1992, and Habitat II, 1996).
• Architectural visions of neotraditional towns and healthy cities (e.g., Van der Ryn and Calthorpe 1986).
• Social capital debate of the 1990s led by Robert Putnam (1995), Amatai Etzioni (1994), and others.[1]
• Application of industrial ecology concepts, environmental audits, and sustainability indicators in the early 1990s by struggling communities attempting to recover from economic downturns and urban decay (e.g., Sustainable Chattanooga).

1. In *The Spirit of Community* (1994) and other works, Amatai Etzioni argues strongly for the idea of communitarianism. His views have provoked considerable debate within the American Sociological Association, with many members arguing that Etzioni has underestimated the importance of achieving economic justice as a precondition for the community values he espouses. Robert Putnam's (1995) controversial work on the decline of social capital—decrying a lack of genuine citizen engagement in collective efforts to improve society—adds another important dimension to the debate, particularly his argument that television has insidiously undermined the role of community in American life.

admired, the techniques of building place—especially town planning—correspond to a culture of community: a commonality based on civic mindedness and social cohesion. Thus for Mumford, early New England meant: (1) the sense of place essential to establishing and maintaining a productive, equitable, and organic relation between built and natural environments; and (2) the sense of the commons, defined by an acceptance of common destiny and by social relations based on dignity and mutual respect. These are the sociocultural characteristics necessary to civic identity in a vibrant regionalist democracy.

Contemporary Influences

Today's sustainable communities movement looks, at first, like an extension of Mumford's vision. It may be more accurate, however, to view it as a reaction to decades of frustration felt by transportation and land-use planners, municipal officials, neighborhood activists, downtown business leaders, and environmental groups faced with their inability to constrain and effectively manage urban sprawl through the use of planning, zoning, and redevelopment tools. Population and economic growth, market forces, housing preferences, and a culture of consumption have heretofore foiled the best of growth management plans. In a pattern repeated across the country, sprawl has consumed land at a rate far in excess of that needed to keep up with population growth. In the Chicago region, for example, land area developed between 1970 and 1990 grew almost fourteen times faster than population growth. During the same period, the number of people living within Los Angeles city limits increased by 45 percent, while land consumption for development and housing increased by nearly 300 percent (Bernstein 1997, 10). California, with 3.5 million acres already in urban use, is consuming over 60,000 acres of open space each year for urban expansion (Fulton 1991, 2). In reaction to these trends, many communities are attempting to manage development in a manner that conserves open space, reduces unnecessary sprawl, and promotes more environmentally benign travel. Beginning with the cities of Ramapo, New York, in the 1960s and Petaluma, California, in the early 1970s, growth control measures were implemented to reduce and in some cases halt the spread of residential, commercial, and industrial development. But few have succeeded. Despite the adoption of hundreds of local growth management ordinances, the overall pattern has been merely to shift development to nearby communities. Hoping to learn from past mistakes, many concerned citizens and local officials have become advocates of Smart Growth—a movement that stresses regional efficiency, environmental protection, and fiscal responsibility in land use decisions. Its supporters include many sustainability advocates.

One of the most visible sources of sustainable community mobilization has been environmentalists, especially those who have lost confidence in the political capacity of national and international institutions to attack environmental problems in a global or comprehensive manner. Practicing what Wildavsky (1979, 41) called a "strategic retreat from objectives," many environmentalists have embraced the 1980s slogan "think globally; act locally," thereby framing and attacking an increasing number of regional, national, and global environmental problems at the local level, where they tend to be more comprehensible, accessible, and tractable. Such thinking received a boost at the 1992 Earth Summit, when participating nations agreed to implement so-called "Agenda 21" plans for both local and national sustainable development. Thus interest in community-oriented strategies has grown as something of an adjunct to sustainable development ideas in general. Critics who dismissed the term *sustainable development* as oxymoronic welcomed the opportunity to replace the word *development* with *community*.

In a similar fashion, scholars, planners, and environmental practitioners who have become dissatisfied with the lack of integration in the management of social and natural systems now embrace sustainable community ideas as a good way to encourage such integration. The sustainable community concept calls for interdisciplinary and crosscutting approaches to problem framing and policy response. It encourages environmentalists to think carefully about the social and economic needs of a community; developers to understand something about ecosystem management; civic leaders to recognize the interdependence of communities in both economic and ecological terms; and ordinary citizens to draw connections between civic engagement and quality of life. As an integration concept, sustainable communities offers a set of mutually enhancing goals for planning and policy making. Elizabeth Kline (1995) argues that sustainable communities represent the successful integration of four key objectives: ecological integrity, economic security, high levels of quality of life, and citizen empowerment with added responsibility.

Competing Orientations

Despite their diversity in aims and background, members of the sustainable community movement share a concern about combining socioeconomic and environmental concerns in such a way as to enhance the well-being of both human and nonhuman beings within a particular place or bioregion. As a matter of emphasis, however, their approaches or orientations can be differentiated by the types of cognitive styles, conceptual and analytical questions, and sustainability indicators that they employ. While some supporters emphasize the technical planning or eco-design features of sustainability (e.g., Aberley 1994), others stress its political and economic determinants (e.g., Pirages 1996, Roseland 1998). Recognizing that there is considerable overlap, it is nevertheless possible to characterize these orientations according to the dominant intellectual foundations and frameworks that supporters use in addressing community sustainability. At least four clusters of sustainability approaches can be identified: (1) the natural capital approach popular with ecological economists; (2) the urban design approach used by many land use planners, architects, and sustainability-minded local officials; (3) the ecosystem management strategy employed by ecologists and many natural resource managers; and (4) the metropolitan governance orientation that characterizes those who seek a regional policy approach to community sustainability. None of these are mutually exclusive, nor can they be said to constitute forms of identity that are commonly or self-consciously accepted by participants in the sustainable communities movement. Box 2.3 presents a summary of the intellectual foundations, challenges, and relevant indicators associated with each orientation.

Capital Theory
Numerous economists and other social scientists have attempted to define sustainability in terms of shared capital or constant capital. For example, Ismail Serageldin (1995), World Bank vice president for environmental programs, has identified the essential requisites of sustainability as the retention for future generations of four types of capital: natural capital, manmade capital, human capital, and social capital. Building on this idea,

Box 2.3
Major approaches to community sustainability

Sustainable community orientation	Underlying foundation	Conceptual challenge	Analytical challenge	Key indicators
Capital theory	economics and accounting	nature as capital	elasticity of subsitution	natural income accounting
Urban design	land-use planning and architecture	planning vs. markets; "design with nature"	constraints posed by existing development	density, open space, traffic flows, jobs/housing balance
Ecosystem management	ecology	systems thinking, natural vs. political boundaries	critical loads and stress points; interconnections	urban habitat type, impervious surface area, size of ecological footprint
Metropolitan governance	regionalism	asymmetrical interdependence	incentives for intergovernmental cooperation	number of regional councils, joint power agreements, metro tax-base sharing

Stephen Viederman (1996, 46) adds a fifth type—cultural capital—as necessary to the mission of sustainable communities:

Sustainability is a community's control and prudent use of all forms of capital—nature's capital, human capital, human-created capital, social capital, and cultural capital—to ensure, to the degree possible, that present and future generations can attain a high degree of economic security and achieve democracy while maintaining the integrity of the ecological systems upon which all life and all production depends.

Viederman's definition, like Serageldin's, is based primarily on an economic notion of sustainability. Although Viederman notes the importance of democracy, the emphasis on capital suggests a concern with instruments of production rather than conditions of exchange. Particularly challenging in this approach is the meaning of *natural* capital and *social* capital. Serageldin (1995, 6) defines natural capital as "the stock of environmentally provided assets (such as soil, atmosphere, forests, water, wetlands) that provide a flow of useful goods and services." Using this concept of capital, neoclassical economists have defined sustainability in terms of nondeclining human welfare over time (Pearce, Markandya, and Barbier 1989). Although such definitions rely entirely on anthropocentric utility, they have the virtue of being quantifiable and monetizable. Of course, some things, such as climate "services," remain extremely difficult to measure in dollars.

Critics both within and outside the field of economics have raised basic concerns about this natural capital approach to sustainability (Stern 1997; Common 1995). A typical objection is that the "constant capital" rule assumes unrealistic degrees of substitution of human-made capital for natural capital. Excessive faith in technology leads to assumptions of high elasticity of substitution, though it is apparent to most observers that many forms of natural capital, such as the ozone layer, have no known artificial substitutes that can be applied on a global scale. Some critics dismiss the notion altogether, arguing that nature's intrinsic value precludes natural capital approaches to sustainability (Spash and Hanley 1995).

Many economists favor natural capital approaches because, at the very least, they comport with widely accepted accounting schemes that permit measurement across disparate domains of value (Jansson et al. 1994). In the case of social capital, however, even crude attempts at measurement may be futile. According to Coleman (1988, S98), this is because "unlike other forms of capital, social capital inheres in the structure of relations between actors and among actors. It is not lodged either in the actors themselves or in physical implements of production." In other words, social capital is about social relations. In terms of sustainable community objectives, social capital is about personal relations that create trust, norms of acceptance, and a local sense of reciprocal obligation. What those relations are, exactly, and how they are developed and maintained remains very difficult to analyze across time and place.

Urban Design

The design orientation appears to dominate the literature on sustainable communities. Architects, land use planners, and civil engineers make up the core support group, but a growing number of government and business leaders have identified this approach as a promising way to revitalize sagging local economies, manage urban growth, and balance demand for transportation, housing, jobs, environmental quality, and social services. Enterprising real estate developers have discovered that projects incorporating sustainability themes may add sales appeal and simplify the entitlement process for securing building permits.

As leaders of the design efforts, architects (e.g., McDonough 1992; Van der Ryn and Calthorpe 1986; Kunstler 1996) have attempted to make communities more pedestrian friendly, energy efficient, conducive to public interaction, and aesthetically attractive. Using labels such as the "New Urbanism," "neotraditional communities," and "Green cities," they have mounted campaigns to redesign neighborhoods and communities along lines similar to those championed by Lewis Mumford more than sixty years ago. Several sets of guiding principles for community design have been prepared as part of these efforts.[7] Among the most important of these design principles is the idea that a community should coexist with nature in a "healthy, supportive, diverse and sustainable condition" (McDonough 1992), and that it should have a well-defined edge, as well as a location, size, and character that permit the close integration of housing, jobs, attractive open space, cultural and recreational amenities, and facilities essential to the daily needs of citizens (Calthorpe 1993).

Although impressed by what these design principles can contribute to the sustainability of new development, many planners and engineers are understandably preoccupied with repairing or restoring the products of past development. The obvious difficulty with most design innovations is that so much of the landscape is already "built out" or settled in ways that appear irreversible and that reflect unsustainable choices. The freedom to select a different course of development or alternative lifestyle is often constrained by the path-dependent outcomes of what has come before (i.e., early design choices have entrained subsequent development and have made it difficult to embark in new directions). Moreover, the economic cost of bold reform, not to mention the political obstacles, render many sustainability designs applicable to only a small number of "boutique" subdivisions, new town demonstrations, or hard-to-replicate experimental communities.

In view of the momentum of past development mistakes, most urban designers point to the importance of infrastructure decisions—especially the allocation of sewer, water, and roads—as fundamental to the success of both new development and redevelopment of existing areas. For many, the design of transportation systems is of paramount importance. David Engwicht (1993), for example, argues that we can sustainably "reclaim" our cities and towns by redesigning transportation on the basis of *exchange* instead of *movement; place* instead of *destination.* The principal challenge becomes one of reducing dependence on private automobiles (Newman and Kenworthy 1989). To that end, design for sustainability usually favors increased density and greater reliance on mixed land uses, both of which encourage walking and greater reliance on public transit.

Ecosystem Management

Although this orientation overlaps significantly with environmentally sensitive forms of urban design, the ecosystem approach can be distinguished chiefly by its emphasis on the role of natural systems, such as watersheds, in providing the physical context for planning and management. In this approach, the preservation and restoration of ecosystems becomes the fundamental consideration in community design and development (Platt, Rowntree, and Muick 1996). Sociopolitical factors have historically been slighted in ecosystem management, although efforts to integrate the human and ecological dimensions of this approach have been strengthened in recent years (Yaffee et al. 1996).

Because it is essentially a science-based management approach to land and other natural resources, ecosystems management requires careful attention to conservation biology, hydrology, and other aspects of field ecology and natural history analysis. Furthermore, because of its costly scientific monitoring requirements and the multi-jurisdictional cooperation it needs in order to succeed, the ecosystem approach usually has to be initiated by federal and state agencies, or large nonprofit environmental organizations, such as the Nature Conservancy (Yaffee et al. 1996). This, and the fact that it utilizes natural boundaries for its planning units, means that ecosystem management often operates in some tension with local political jurisdictions. It is by definition a regional approach to sustainability, and thus shares a key feature of the metropolitan approach discussed below.

Bioregionalism, a less science-based ecosystem orientation, has proven very popular with some sustainable community advocates in Europe and North America. Their aim has been to develop a nontechnical understanding of ecological interdependence among ordinary citizens who share a particular bioregion, thus helping the residents recognize their personal stake in the proper management of ecosystems. Relying more on the insights of Deep Ecology than on scientific research, bioregionalists seek to define human communities as subsystems of ecological communities (Naess 1989). This poses the obvious challenge of reconceptualizing one's place in terms of biophysical features and their interconnections, rather than according to human settlement features, such as roads, towers, reservoirs, and the usual landmarks of the built environment. Moreover, it changes the human role with regard to nature from that of owner to steward or mere tenant. As one leading advocate of bioregionalism sees it, the challenge is to become "dwellers in the land," interacting in a society whose scale of community, government, and economy is small enough to permit intimate integration with the natural world (Sale 1985).

A hybrid of the ecosystem and urban design approaches, applying the characteristics of healthy natural systems to human communities, has been developed by the Chicago-based Urban Sustainability Learning

Group (1996). According to their view, a sustainable community is constantly changing through continual interdependent fluctuations in order to maintain itself as the environment changes.

It has institutions that are loosely coupled, decentralized and open to new members, and have shifting and overlapping membership. It has multiple elements and pathways that fulfill its functions, and it facilitates the efforts of its members to participate and develop unique niches. It is interdependent through a vast network of loose relationships among its people and institutions and between the whole economy and other parts of the environment. The network is supported by easy and continuous trading of information and goods. It has feedback loops that allow it to self-regulate . . . and it has some common values about desired community and environmental conditions. (33)

Metropolitan Governance

A simpler but more politically oriented conception of sustainable communities might define them as communities that achieve and retain improvements in quality of life without diminishing the quality of life enjoyed by other communities, now and in the future. Such a definition calls ideally for a global and intertemporal sense of human responsibility. Realistically, of course, parochialism and contempocentrism (a preoccupation with the needs and opportunities of the present) remain major impediments to development of a planetary sense of stewardship and social concern for future, nonoverlapping generations. The best that can be expected, perhaps, is for members of a community to adopt a regional perspective.

The zero-sum notion of sustainability implied in this definition requires that communities avoid strategies that achieve local sustainability at the expense of other communities. By importing more than their fair share of critical resources and exporting their waste and pollution, a community can improve the quality of life of its citizens by externalizing the costs of improvement, both economically and geographically. Clearly, making air quality improvements in Los Angeles "sustainable" by increasing the city's reliance on coal-fired power plants in Utah is not what community sustainability is about. On the other hand, given differential resources, capabilities, and needs, how can competing communities avoid a certain amount of "beggar-thy-neighbor" behavior? And what constitutes a "fair share" of critical resources among communities anyway?

Given that today's community needs and opportunities can be heavily influenced by international trade arrangements and foreign investment, it is very difficult to devise allocation rules that work to the benefit of both individual communities and regions (or global systems) as a whole.

Because of the enormous expansion of economic trade and technology, and in some cases the legacy of colonialism, the locations of a community's economic activity and ecological impact no longer coincide with its geographic location (Wackernagel and Rees 1996). In most countries, communities increasingly appropriate carrying capacity from other communities and regions. The ecological services of one bioregion are used to subsidize the growth of human activities in another. Imported ecological wealth and exported waste permit communities to hide the true costs of unsustainable practices, externalizing them in distant places in the form of huge "ecological footprints."[8] In sprawling Los Angeles County, for example, the area of ecologically productive land needed to sustain the lifestyles of over 9 million residents is nearly 350,000 square miles, more than eighty-five times larger than the county itself (Venetoulis and Hempel 1998, 3). To say that a community can become sustainable by externalizing the costs of its waste and consumption is to misconstrue the meaning of sustainability.

The point of this "individualistic" account of communities is that sustainability may require something different: a "community-of-communities" approach. The practical implication is that sustainable communities may have to be assessed regionally, if not globally. It is the interactions between communities as much as the interactions within, that determine each community's potential for sustainability. The political implication of this argument is a renewed emphasis on regionalism and what I have termed glocalism (Hempel 1996a).[9] Spurred in part by the continuing devolution of power from the federal government to state and local levels, and by the increasing globalization of key economic sectors, the need to develop and connect global and local policy making is becoming apparent to all but the most reactionary followers of nationalism.

Because the strong interdependence of communities is seldom reflected in the coordination mechanisms and intergovernmental cooperation agreements that exist today, a regional framework of governance may be necessary for community-based concepts of sustainability to succeed.

Without a regional vision and the institutional arrangements to facilitate it, the goals of sustainable community end up looking parochial and selfish. This does not mean that sustainability requires formal institutions of regional government per se. It does, however, suggest the need to create incentives for greater regional and subregional cooperation in promoting environmental quality and economic opportunity, while discouraging trends and developments (e.g., excessive growth in population, resource consumption, and undesirable land uses) that strain regional capacity for ecological renewal and for political representation, coordination, and cooperation.

Given that over 75 percent of Americans now live in one of 320 metropolitan regions, the "community-of-communities" idea appears to be less far-fetched than our fragmented political system of more than 80,000 local jurisdictions would suggest.[10] As Richard Levine argued so succinctly (Hiss 1996, 7), metropolitan regions may constitute "the largest unit capable of addressing the many urban, architectural, social, economic, political, natural resource, and environmental imbalances in the modern world, and, at the same time, the smallest scale at which such problems can be meaningfully resolved in an integrated and holistic fashion."

Two distinct and opposing political strategies find expression among supporters of metropolitanism. The first, promoted by business and civic leaders, emphasizes elite consensus building and reliance on public-private partnerships to build regional cooperation. Neal Pierce, a leading advocate of this strategy, argues that regional approaches offer the only effective way to attack crosscutting problems of urban environmental quality, economic development, transportation access and mobility, and the growing geographic inequities caused by what Robert Reich terms the "secession of the successful" into wealthy suburbs. For supporters of this strategy, the key to success lies in getting the rich and powerful to agree on regional solutions.

But is regionalism likely to arise out of such consensus-building efforts? The idea of promoting sustainable communities through regional cooperation is attractive, but the stakeholder politics involved are clearly daunting.[11] In response, an alternative and more provocative view has been injected into the national debate by Myron Orfield (1997), who proposes the use of majority coalition politics to bring about redistribution of regional resources and opportunities. Convinced that consensus-building efforts usually lead to outcomes that favor wealthy communities, Orfield favors a more confrontational route. Focusing initially on the Twin Cities (Minneapolis-St. Paul) region of Minnesota, Orfield has attempted to harness intraregional polarization in the interest of equity and sustainability. This polarization, some of which is latent, arises from highly skewed tax bases, social services, and competition for economic development that combine to undermine local cooperation. Concentrated poverty is one of the most disturbing symptoms, but distressed neighborhoods and subregions also exhibit serious problems of environmental deterioration, poor schools, violent crime, and political isolation. Without writing explicitly about sustainable community concepts, Orfield nevertheless lays out a political strategy for operationalizing a number of sustainability ideas across metropolitan areas. The political core of his strategy lies in building coalitions between inner city neighborhoods and blue-collar suburbs for the purpose of forcing more equitable tax base sharing with wealthy suburbs, thereby helping to redirect urban expenditures toward a more balanced and integrated portfolio of environmental, social, and economic investments, spread across diverse communities. Like the urban design approach, his strategy treats infrastructure decisions as the key leverage points for success. Transportation, energy, water, and sewer systems represent the "visible hand" of community design, and decisions about their provision factor heavily in his view of progress toward sustainability.

Operationalizing the Concept

As the previous discussion suggests, notions of sustainability frequently trade on lofty sentiments and principles that are difficult, if not impossible, to implement one community at a time. Furthermore, low levels of civic engagement and ecological literacy make it difficult to comprehend what is at stake in choices about regional governance, urban design, ecosystem management, and capital substitution. Most discussion about sustainable communities takes place at the "high-concept" level, rather than in a context of practical problem solving. As a result, efforts to incorporate sustainability goals into public policy may be easy at an abstract level, but once a specific link is made to energy consumption, transportation, land use, population growth, air quality, economic development, and so forth, the utility of the concept begins to erode. Not only is the idea hard to grasp intellectually—because it requires synthetic thinking and integrative minds—but it is also hard to measure in ways that illuminate the concept.

Community Indicators
One of the most promising ways to move from "clouds to concrete" is to construct reliable sustainability indicators that can be tailored to diverse communities and periodically monitored for changes in direction or intensity. These indicators of sustainability measure regional and community-level capacity to meet present and future needs in ways that do not sacrifice important opportunities for future generations. They are essentially integrative measures of economic, social, and ecological health that are designed to gauge a community's systemic balance and resilience over long periods of time. Examples of indicators (see boxes 2.4 and 2.5) include the ratio of job growth to population growth, pounds of solid waste landfilled per capita, high school graduation rates, and the greenspace-blackspace ratio (open space:paved area). No single indicator is adequate to measure a community's sustainability, but taken in combination a carefully selected battery of indicators can reveal much about a community's movement toward or away from sustainability goals.

More sophisticated sustainability indicators can be used to measure everything from institutional capacity (e.g., annual tax capacity per household to support community services) to noise exposure (e.g., number of children exposed to air and ground traffic noise levels above 50 decibels from 8 p.m. to 8 a.m.). Using advanced information technologies—especially geographic information systems (GIS)—these indicators can be overlaid, mapped, and compared within and across regions to provide sophisticated ways to visualize the complex interactions that influence a community's sustainability index or "scorecard."

Geographic information systems provide a practical tool for analyzing how communities are changing over time and why. By integrating layers of information about social, environmental, and economic sustainability, GIS users can produce a color-keyed map of trends and driving forces

Box 2.4
Examples of sustainable community indicators

> - Percent of workforce concentrated in largest five employers
> - Greenhouse gas emissions per capita
> - Number of domestic violence calls to police
> - Number of community gardens
> - Landfilled solid waste (tons per year)
> - Percent of households that can afford median-priced house
> - Water use and wastewater flows (gallons/day/person)
> - Net growth in livable wage jobs
> - Graduation rate by race and ethnicity
> - Energy consumption (BTU per capita)
> - Voter turnout in municipal elections
> - Pounds of toxics produced and released per year
> - Dollar value of repairs or replacement needed in infrastructure investment
> - Homeownership rate
> - Number of endangered and threatened species
> - Community volunteerism by age group (survey data)

that interact spatially at the neighborhood, community, and regional levels. Although universities, corporations, government agencies, and planning organizations have been using GIS for more than a decade, only a few applications have been developed thus far for the purpose of tracking sustainability indicators. The recent introduction of simple yet highly sophisticated desktop GIS software promises to make this capability available and affordable to even the smallest local governments, school districts, and community organizations.

Not surprisingly, while the capability to construct, monitor, and visualize complex sustainability indicators has increased markedly in the past decade, the utilization of such information in policy making has been rather limited. Political sensitivity about what indicators reveal is not the least of the reasons for this relative lack of progress. Other considerations have also been important: problems of unreliable, invalid, or missing data; problems of data that are not available over time, or not comparable for time series analysis; interjurisdictional conflicts in data collection and interpretation; indicators that are unrepresentative for policy making, or irrelevant to important community goals; indicators that raise

Box 2.5
Three types of community indicators

General	Quality of life	Sustainability
vehicle miles traveled	congestion levels	total vehicular CO_2 emissions/year
economic growth rate	income/capita	ratio of income/capita to municipal debt/capita
population growth rate	urban density levels	ecological footprint (productive land appropriated)
solid waste (tons/yr)	solid waste/capita	percent of solid waste diverted to recycling
number of parks	park acreage/1,000 people	ratio of greenspace-blackspace (paved area)
size of police force	violent crimes/10,000 people	alcohol and drug treatment beds/10,000 people
housing supply	median housing price	percent of adults that can afford median home
job growth rate	unemployment rate	percent of new jobs paying a livable wage
business permits	business vacancy rate	ratio of business startups to business failures
resource consumption	water shortages/year	ratio of renewable water supply to withdrawals
health facilities	health insurance coverage (%)	packs of cigarettes sold per person per year
school facilities	student-teacher ratios	number of courses promoting ecological literacy

privacy concerns; and indicators that are too complex or esoteric to be understandable by decision makers, let alone the general public. In some instances, the sheer growth of indicators has overwhelmed human capacities to process and interpret the information they contain. The Alberta Round Table on Environment and Economy, for example, initially developed a list of 850 potential sustainability indicators before coming up with a more manageable list of 59 (Maclaren 1996, 197).

Beyond the data analysis and management limitations lies one more critical constraint on the development and use of sustainability indicators: the role of ordinary citizens in their selection and interpretation. Deliberative democracy is, to many people, both a means and an end of the sustainable community movement. If deliberative democracy is conducive to the process of sustainability, and vice versa, it is important that citizens participate in the selection of indicators that will be used to evaluate their community and region. Although such involvement will sometimes lead to the inclusion of indicators that scholars and professional analysts regard as unscientific, irrelevant, or unreliable,[12] to exclude such involvement may reveal, as clearly as any indicator, a basic cause of unsustainability—lack of civic engagement.

Conclusion

Deciding which sustainability approaches and indicators to adopt requires a judicious balancing of technical merit and popular appeal. The capital theory, urban design, ecosystem management, and metropolitan governance approaches all offer important perspectives and insights for use in becoming more sustainable. Each has its limitations as well. Although elements of all four approaches can be included in sustainability strategies, it is apparent that some approaches will have more public appeal than others. For example, urban design is likely to attract more adherents than ecosystem management, and both will have more appeal than capital theory, with its dry academic tone. Moreover, each approach, along with the indicators employed, has to be custom-tailored to suit the needs of diverse communities with unique problems and opportunities.

One perspective in particular—metropolitan governance—deserves special attention as a means for developing the regional policy framework needed to make the other approaches more feasible and effective. For many communities, improved governance appears to be a prerequisite for progress in sustainability. The politics of the metropolitan approach, in combination with the science of the ecosystem approach, may offer the best hope for resolving some of the problems of parochialism and unplanned growth that threaten the long-term sustainability of these communities. By emphasizing the importance of intercommunity cooperation in achieving sustainability, the metropolitan approach can in theory help to embed the other approaches in the essential workings of local politics and regional policy formation.

In practice, of course, metropolitanism has not fared well in American politics. Like Deborah Stone's (1997) model of society based on the polis rather than the marketplace, the metropolitan vision rests on a little used model of democratic regionalism, rather than on the more familiar models of local control, consumer markets, and design by experts. As such, it promotes the ideal of a community of communities, while at the same time attempting to limit the freedom of individual communities to pursue designs, development strategies, and capital allocations that are unsustainable for the region as a whole.

While arguments can be made that such an approach merely shifts the struggle for competitive advantage in growth and consumption from individuals to communities and on to the regional (subnational) level, it appears that most of today's long-term threats to sustainability are more visible and, arguably, more tractable at this level—that is, regional policies may hold greater promise than those arising at the household, local, national, multinational, or global levels. Regionalism, however, is not a panacea. Metropolitan economies could end up competing with one another in ways that are just as debilitating and unsustainable as the competition that now occurs between many central cities and their suburbs. Fortunately, competition among communities and regions need not be (and seldom is) "zero-sum." What makes the metropolitan approach particularly promising for building sustainable communities is its scale of action and exchange: it is big enough to capture key ecological, social, and economic interdependencies, yet small enough to provide a sense of place and social embeddedness. In theory, it affords an optimal scale at which to attempt the integration of governance, social diversity, economic development, and environmental protection.

Eventually it will be necessary to think and act globally in an effort to become more sustainable. Metropolitan approaches may become less effective or attractive than state programs, larger regional approaches, federal initiatives, or planetary solutions. For the time being, however, it is at the metropolitan level that sustainability concepts appear most operational and meaningful. This is the level where human communities interact most tellingly with nonhuman nature. It is the primary arena in which conflicting ideals of community, liberty, and ecology will have to be reconciled with the forces of economic globalization.

It remains to be seen if the American political system will adjust to this kind of thinking. The goals of sustainable community may be too Jeffersonian to suit a society shaped predominantly by Hamiltonian precepts. Then again, they may provide an important stimulus for rethinking what quality of life requires in the new millennium.

In *Democracy in America* (1835), de Toqueville observes that among the key enabling features of American politics are "those township institutions which limit the despotism of the majority and at the same time impart to the people a taste for freedom and the art of being free" (Bradley 1945, I, 288). A sustainable community is one in which such "taste for freedom" is derived ecologically, as well as politically. Without an ecological reconception of freedom as sustainability, we are left to ponder how and under what conditions, if any, mortal political actors, motivated by short-term incentive structures, will choose to adopt policies that promote genuine sustainability.

Notes

1. The term *sustainable community* has many variations, including *green cities, eco-cities, sustainable cities, eco-communities,* and *livable cities.* The sustainable communities literature has been described by Roseland (1994) as consisting of at least ten distinct streams of thought that bring together urban and community-based designers, practitioners, visionaries, and activists. The sources of many of these streams can be traced to the literature on urban ecology, growth management, bioregionalism, neotraditional architecture, environmental planning, community self-reliance, and appropriate technology.

2. See, for example, Pirages 1996; Hart 1997; Schmidheiny 1992; the President's Council on Sustainable Development 1996; Milbrath 1991; Hawken 1993; and Common 1995.

3. Common (1995) notes that systems with low stability—for example, temperate forests—can still have high resilience; just as highly stable systems—for example, tropical ecosystems—can have low resilience compared with temperate ones.

4. Although many writers draw a distinction between development and growth—with growth (interpreted as "bigger") being ecologically unsustainable, unlike development (interpreted as "better")—some critics argue that "development cannot be purified of its historical context" (Sachs 1993: 21) as both a means and end of growth.

5. Of particular relevance here are the arguments made by members of the International Forum on Globalization. Prominent spokespersons include Vandana Shiva, Martin Khor, David Korten, Jerry Mander, Wolfgang Sachs, and Helena Norberg-Hodge.

6. A listing of these sites is available both by topic and alphabetically at the Web Site Link Library of the Global Environmental Options Network: www.geonetwork.org/links/aboutlink.html. Three additional and very useful sustainable community link sites are www.sustainable.doe.gov/, www.rprogress.org/, and www.subjectmatters.com/indicators.

7. Perhaps the best known of these guides are the Hannover Principles, developed by William McDonough, dean of the School of Architecture at the University of Virginia, and the Ahwahnee Principles, developed in 1991 by Peter Calthorpe and others at the instigation of the California Local Government Commission.

8. Quantifying the total appropriated carrying capacity of a community is extremely difficult, owing in part to conceptual difficulties with carrying capacity itself (Cohen, 1995, 237–260). But such quantification is nevertheless important for those who wish to operationalize the concept of sustainability. A promising though crude measure for this purpose—ecological footprint analysis—has been developed by professor William Rees of the University of British Columbia and his former student Mathis Wackernagel. Wackernagel and Rees (1996) conceive of footprints as accounting tools for calculating human impacts in terms of the land and water areas appropriated for energy and resource consumption, and for waste disposal. An ecological footprint is a measure of the load placed on the biosphere by a given population. Footprints are therefore proportional to a community's combined population and per capita consumption levels.

9. An apparent irony of sustainable community thinking is that it may require regional or, ideally, global orientations toward urban quality of life issues. I have termed the latter orientation "glocal," owing to its concern with both global economic opportunity and local political capacity, and their linkage through governance. The rationale for glocalism is derived from the need to connect local policy making processes that are conducive to deliberative democracy with sustainability strategies that must sometimes be global, or at least regional, in order to be effective. Re-establishing the primacy of community in political life may facilitate the social and environmental sensibilities needed to manage the increasingly global reach of technology and capital.

10. Included in local political jurisdictions in the United States are approximately 3,000 counties, more than 19,000 municipalities, over 16,000 townships, nearly 15,000 school districts, and roughly 30,000 special districts.

11. The fiscal implications of the increasing concentration of wealth and poverty in different parts of metropolitan areas represent major obstacles to the goals of sustainable community. The political power of affluent suburbs has increased, and so has their resistance to regional approaches that involve metropolitanwide responses, such as tax base sharing. The 1992 presidential election marked the first time in history when suburban voters constituted an absolute majority. After redistricting in 1992, suburban districts in the House outnumbered urban districts by 170 to 98—the rest being rural and mixed urban-suburban or rural-suburban (Dreier 1996).

12. For example, the author was recently involved in a sustainable community indicator project in which local residents insisted that measures of graffiti removal would provide one of the best sustainability indicators for their neighborhood. Some academic members of the project team regarded the local residents' ideas as shortsighted, crude, or trivial, preferring instead measures that could be derived from the scientific literature.

References

Aberley, Doug. ed. 1994. *Futures by Design: The Practice of Ecological Planning.* Philadelphia, PA: New Society Publishers.

Berman, Marshall. 1982. *All that Is Solid Melts into Air.* New York: Penguin.

Bernstein, Scott, 1997. "Community-Based Regionalism Key for Sustainable Future" *The Neighborhood Works* 20,6 (November–December):10.

Bradley, Phillips, ed. 1945. Alexis de Tocqueville's *Democracy in America.* Vol. 1. New York: Vintage.

Brundtland Commission (World Commission on Environment and Development). 1987. *Our Common Future.* New York: Oxford University Press.

Calthorpe, Peter. 1993. *The Next American Metropolis: Ecology, Community, and the American Dream.* Princeton, NJ: Princeton Architectural Press.

Campbell, Scott. 1996. "Green Cities, Growing Cities, Just Cities?—Urban Planning and the Contradictions of Sustainable Development." *Journal of the American Planning Association* 62,3:296–311.

Cohen, Joel. 1995. *How Many People Can the Earth Support?* New York: W. W. Norton.

Coleman, James S. 1988. "Social Capital in the Creation of Human Capital." *American Journal of Sociology (Supplement)* 94:S95–S120.

Common, Michael. 1995. *Sustainability and Policy: Limits to Economics.* Cambridge University Press.

Dovers, Stephen R. 1997. "Sustainability: Demands on Policy." *Journal of Public Policy* 16,3:303–318.

Downs, Anthony. 1994. *New Visions for Metropolitan America*. Washington, DC: The Brookings Institution, and Massachusetts: Lincoln Institute of Land Policy.

Dreier, Peter. 1996. "The Struggle for Our Cities: Putting the Urban Crisis on the National Agenda." Paper prepared for the 1996 Planners Network Conference, "Renewing Hope, Restoring Vision: Progressive Planning in Our Communities."

Engwicht, David. 1993. *Reclaiming Our Cities and Towns: Better Living With Less Traffic*. Gabriola Island, BC: New Society Publishers.

Etzioni, Amatai. 1994. *The Spirit of Community: The Reinvention of American Society*. New York: Simon and Schuster.

Euston, Stanley, 1995. "Gathering Hope: A Citizens' Call to a Sustainability Ethic for Guiding Public Life." Sante Fe, NM: The Sustainability Project.

Folke, Carl, and Tomas Kaberger. 1991. "Recent Trends in Linking the Natural Environment and the Economy." In *Linking the Natural Environment and the Economy: Essays from the Eco-Eco Group,* Folke and Kaberger, eds. (Dordrecht: Kluwer).

Fuller, R. Buckminster. 1969. *Operating Manual for Spaceship Earth*. New York: Simon and Schuster.

Fulton, William. 1991. *Guide to California Planning*. California: Solano Press Books.

Geddes, Patrick. 1915. *Cities in Evolution*. New York: Harper and Row.

Hart, Stuart L. 1997. "Beyond Greening: Strategies for a Sustainable World." *Harvard Business Review* 75,1 (January-February):66–76.

Hawken, Paul. 1993. *The Ecology of Commerce: A Declaration of Sustainability*. New York: HarperCollins.

Hempel, Lamont C. 1998. *Sustainable Communities: From Vision to Action*. Claremont Graduate University.

Hempel, Lamont C. 1996a. *Environmental Governance: The Global Challenge*. Washington, DC: Island Press.

Hempel, Lamont C. 1996b. "Roots and Wings: Building Sustainable Communities." White Paper, League of Women Voters Population Coalition (January).

Hempel, Lamont C. 1992. "Earth Summit or Abyss?" Paper presented at the Global Forum, United Nations Conference on Environment and Development, Rio de Janeiro, Brazil (June).

Hiss, Tony, 1996. "Outlining the New Metropolitan Initiative." (Project Report) Chicago: Center for Neighborhood Technology.

Holling, C. S. 1973. "Resilience and Stability of Ecological Systems." *Annual Review of Ecology and Systematics* 4:1–24.

Howard, Ebenezer. 1898. *Tomorrow: A Peaceful Path to Real Reform.* Later published as *Garden Cities of Tomorrow.* Cambridge, MA: MIT Press, 1965.

International Union for Conservation of Nature (IUCN), United Nations Environment Program and World Wildlife Fund. 1980. *World Conservation Strategy.* Gland, Switzerland: IUCN.

Jansson, Ann-Marie, Monica Hammer, Carl Folke, and Robert Costanza. 1994. *Investing in Natural Capital: The Ecological Economics Approach to Sustainability.* Washington, DC: Island Press.

Kline, Elizabeth. 1995. *Sustainable Community Indicators: Examples from Cambridge.* Boston, MA: Tufts University Consortium for Regional Sustainability.

Kunstler, James Howard. 1996. *Home from Nowhere: Remaking Our Everyday World for the Twenty-First Century.* New York: Simon and Schuster.

Luccarelli, Mark. 1995. *Lewis Mumford and the Ecological Region: The Politics of Planning.* New York: The Guilford Press.

Maclaren, Virginia W. 1996. "Urban Sustainability Reporting." *Journal of the American Planning Association* 62,2: 184–202.

McDonough, William. 1992. "The Hannover Principles: Designing for Sustainability." Report to the City of Hannover, Germany (a guide for international design competitions for EXPO 2000).

McHarg, Ian. 1969. *Design with Nature.* Garden City, NY: Natural History Press.

Meadows, Donella H., Dennis L. Meadows, Jorgen Randers, and William Behrens III. 1972. *The Limits to Growth.* New York: Universe.

Milbrath, Lester. 1991. *Envisioning a Sustainable Society.* Albany: State University of New York Press.

Morris, David. 1982. *Self-Reliant Cities: Energy and the Transformation of Urban America.* San Francisco: Sierra Club Books.

Mumford, Lewis. 1964. *The Myth of the Machine.* New York: Harcourt Brace Jovanovich.

Mumford, Lewis. 1926. *The Golden Day: A Study in American Experience and Culture.* New York: Boni and Liveright.

Mumford, Lewis. 1924. *Sticks and Stones: A Study of American Architecture and Civilization.* New York: Boni and Liveright.

Munro, David A. 1995. "Sustainability: Rhetoric or Reality?" In *Sustainable World: Defining and Measuring Sustainable Development,* ed. Thaddeus C. Tryzyna. Sacramento, CA: International Center for the Environment and Public Policy and the World Conservation Union: 27–35.

Naess, Arne. 1989. *Ecology, Community and Lifestyle* (Translated and edited by David Rothenberg). New York: Cambridge University Press.

Newman, Peter, and J. R. Kenworthy. 1989. *Cities and Automobile Dependence.* Brookfield, VT: Gower Technical.

Ophuls, William. 1997. *Requiem for Modern Politics: The Tragedy of the Enlightenment and the Challenge of the New Millennium.* Boulder, CO: Westview Press.

Ophuls, William. 1996. "Unsustainable Liberty, Sustainable Freedom." In *Building Sustainable Societies: A Blueprint for a Post-Industrial World,* ed. Dennis Pirages. Armonk, NY: M.E. Sharpe.

Orfield, Myron. 1997. *Metropolitics: A Regional Agenda for Community and Stability.* Washington, DC: Brookings Institution/Cambridge: Lincoln Land Institute.

Pearce, David W., Anil Markandya, and Edward B. Barbier. 1989. *Blueprint for a Green Economy.* London: Earthscan.

Pirages, Dennis. 1996. *Building Sustainable Societies: A Blueprint for a Post-Industrial World.* New York: M.E. Sharpe.

Platt, Rutherford H., Rowan A. Rowntree, Pamela C. Muick, eds. 1996. *The Ecological City: Preserving and Restoring Urban Biodiversity.* Amerst, MA: University of Massachusetts Press.

President's Council on Sustainable Development. 1996. *Sustainable America: A New Consensus for Prosperity, Opportunity, and a Healthy Environment for the Future.* Washington, DC: US Government Printing Office.

Putnam, Robert D. 1995. "Tuning In, Tuning Out: The Strange Disappearance of Social Capital in America." *PS: Political Science and Politics* (December):664–683.

Rees, William E. 1990. "Economics, Ecology and the Role of Environmental Assessment in Achieving Sustainable Development." In *Sustainable Development and Environmental Assessment: Perspectives on Planning for a Common Future,* eds. P. Jacobs and B. Sadler. Ottawa: CEARC-FEARO.

Register, Richard. 1987. *Eco-City Berkeley: Building Cities for a Healthy Future.* Berkeley, CA: North Atlantic Books.

Roseland, Mark, 1998. *Toward Sustainable Communities: Resources for Citizens and their Governments.* Gabriola Island BC, Canada: New Society Publishers, 2d ed.

Roseland, Mark. 1994. "Sustainable Communities: An Examination of the Literature." In *Sustainable Communities Resource Package.* Toronto: Ontario Round Table on the Environment and the Economy (April 1995).

Sachs, Wolfgang, ed. 1993. *Global Ecology: A New Arena of Political Conflict.* London: Zed Books.

Sale, Kirkpatrick. 1985. *Dwellers in the Land: The Bioregional Vision.* San Francisco: Sierra Club Books.

Schmidheiny, Stephan. 1992. *Changing Course: A Global Business Perspective on Development and the Environment.* Cambridge, MA: MIT Press.

Serageldin, Ismail. 1995. "Sustainability and the Wealth of Nations: First Steps in an Ongoing Journey (Draft)." Paper presented at the Third Annual World

Bank Conference on Environmentally Sustainable Development. Washington, DC.

Spash, Clive L., and Nick D. Hanley. 1995. "Preferences, Information, and Biodiversity Preservation." *Ecological Economics* 12: 191–208.

Stern, Andrew. 1997. "Capital Theory Approaches to Sustainability: A Critique." *Journal of Economic Issues* (March).

Sussman, Carl, ed. 1976. *Planning the Fourth Migration: The Neglected Vision of the Regional Planning Association of America.* Cambridge, MA: MIT Press.

Toffler, Alvin. 1970. *Future Shock.* New York: Random House.

Treanor, Paul. 1996. "Why Sustainability Is Wrong." Electronic publication available at webinter.nl.net/users/Paul.Treanor/sustainability.html.

Urban Sustainability Learning Group. 1996. *Staying in the Game: Exploring Options for Urban Sustainability.* Available from the Center for Neighborhood Technology, Chicago, IL.

Van der Ryn, Sym, and Peter Calthorpe. 1986. *Sustainable Communities: A New Design Synthesis for Cities, Suburbs and Towns.* San Francisco: Sierra Club Books.

Venetoulis, Jason, and Lamont Hempel. 1998. "Los Angeles County's Footprint: Prospects for Sustainability."(Project report) Claremont Graduate University.

Viederman, Stephen. 1996. "Sustainability's Five Capitals and Three Pillars." In *Building Sustainable Societies: A Blueprint for a Post-Industrial World,* ed. Dennis Pirages. Armonk, NY: M.E. Sharpe.

Voinov, Alexey and Courtland Smith. 1998. "Dimensions of Sustainability." Electronic publication available at http://kabir.cbl.cees.edu/AV/PUBS/DS/Sust_Dim.html.

Wackernagel, Mathis. 1997. Personal communication (November).

Wackernagel, Mathis, and William Rees. 1996. *Our Ecological Footprint: Reducing Human Impact on the Earth.* Philadelphia: New Society Publishers.

Wildavsky, Aaron. 1979. *Speaking Truth to Power: The Art and Craft of Policy Analysis.* Boston: Little, Brown & Company.

World Commission on Environment and Development. 1987. *Our Common Future.* Oxford, UK: Oxford University Press.

Yaffe, Steven, A. F. Phillips, I. C. Frentz, P. W. Hardy, S. M. Maleki, B. E. Thorpe. 1996. *Ecosystem Management in the United States: An Assessment of Current Experience.* Washington, DC: Island Press.

II

Transitional Approaches in Conventional Media-Based Environmental Policies

3

Los Angeles' Transition from Command-and-Control to Market-Based Clear Air Policy Strategies and Implementation

Daniel A. Mazmanian

Cleaning the nation's air has been at the top of the agenda throughout the modern environmental movement. The most important legislation to this effort has been the Clean Air Act of 1970, in which Congress directed the newly established U.S. Environmental Protection Agency (EPA) to set national air quality standards to protect the public health. These National Ambient Air Quality Standards were initially set for five criteria pollutants—carbon monoxide, particulate matter, nitrogen oxide, ozone, nitrogen dioxide, and sulfur dioxide. Lead was later added to the list as was small particulates in 1987 and then even smaller "fine" particulates in 1997. Implementing air pollution controls for all industry and business across the nation as well as for motorized vehicles was not going to be an easy task under the best of circumstances. This combined with the choir of complaints that arose from state and local governments, business, and industry over EPA's initial implementation effort, led in 1977 to Congress's amending the Clean Air Act. Congress shifted to the states and in some instances local and regional air pollution control agencies partial responsibility for the Act's implementation (Kamieniecki and Ferrall 1991; Bryner 1993).

This top-down, federal-state-local government regulatory approach has been notably successful. For the nation as a whole, lead emissions shrank by 98 percent between 1970 and 1995, largely due the government's mandating of lead-free gasoline. During the same period, emissions of small particulates from industry and fuel combustion declined by nearly 80 percent, sulfur dioxide by 40 percent, carbon monoxide by 28 percent, and ozone by 35 percent. The link between emissions of criteria pollutants and human exposure and health risk, however, is not exact. Local variations in emissions, weather patterns, differences in lifestyles and individual susceptibility all make a difference. What is certain is that even after twenty-five years of progress in emission reductions, in 1995 the EPA believed that at least 80 million people were still living in "nonattainment areas," counties where one or more of the emissions criteria was not being met. Thus, while the first quarter century of the modern environmental movement saw impressive improvement in air quality, further reduction is required in a number of communities across the nation (Davies and Mazurek 1998, ch. 5).

That Los Angeles has had the unenviable distinction of being the most heavily polluted metropolitan air basin in the nation and continues as the most serious nonattainment region today is well known. Less well known is that there has been a nearly unbroken decline in health alerts and emissions violations in the region, possibly more dramatic in relative terms than in any other metropolitan area in the nation. The significant (albeit still insufficient) decrease in air pollution in Los Angeles is impressive because success has been achieved despite the fact that it has been one of the most rapidly expanding urban centers in the nation, with unprecedented economic growth, and despite seemingly endless urban sprawl (see box 3.1). So while some see the cup as half empty, others can see it as half full. Possibly least recognized of all, the implementation of clean air policy in Los Angeles during the first environmental epoch had been carried out by the largest, most encompassing, regulatory, command-and-control, regional air pollution control regime in the nation.

Also, this regime was one of the first to be seriously challenged to reengineer itself and refocus its activities at the onset of the second environmental epoch, to bring it more in line with changing political values and policy approaches. The regime would need to undertake a number of fundamental changes to simultaneously survive organizationally in the new political environment and maintain the downward trend in pollution reduction that people in the region had come to expect.

This chapter traces the rise of the regulatory regime in Los Angeles— which encompasses not just the city and county of Los Angeles but also the adjacent counties of Orange, Riverside, and the western section of San Bernardino—and its retreat and partial replacement during the second environmental epoch. In the context of the general themes of the

Box 3.1
Noteworthy facts about Los Angeles

• From 1950 through the mid-1990s, the population in the Los Angeles basin has more than tripled, from 4.8 million to nearly 16 million people (while California has grown to more than 32 million. Although the rate of growth in the early 1990s slowed appreciably, current predictions are that the state will increase by another 25 percent by 2020.

• The number of motor vehicles on the road has more than quadrupled, from 2.3 million to 10 million from the 1940s through the mid-1990s.

• The state's economy is ranked between the seventh to eighth largest, compared with all nations of the world, and the Los Angeles region itself, tenth or eleventh. Until recently the region was one of the nation's major aerospace and defense centers and it remains extremely active in textiles and design, light manufacturing, film making, media, software and electronics, computing, and finance. The combined ports of Los Angeles and Long Beach (which are adjacent) lead the nation in volume of imports and trade.

• The region enjoys a mild climate most of the year, with ocean breezes blowing in from west to east at an average of 5 miles per hour. For all the region's problems, inhabitants still enjoy the sun, the beaches, and the surrounding mountains.

• The South Coast Air Quality Management District covers approximately 12,000 square miles, essentially the Los Angeles basin—Los Angeles, San Bernardino, Riverside, and Orange counties—and has planning jurisdiction over the desert beyond. It is bounded by the Pacific Ocean on the west, and the San Gabriel, San Bernardino, and San Jacinto mountains to the north and east.

• The region remains the most severely polluted, major urban air shed in the nation, violating NAAQS somewhere in the basin one out of every three days of the year. It is the only place classified by EPA as "extreme" in air pollution.

• During the first half of the 1990s, the Los Angeles region was hit the hardest of any region in the country by the demise of defense spending and the aerospace industry, with 400,000 lost jobs, a 30 percent decline in property values, an exodus of skilled workers, and a slowdown in population growth.

• The Los Angeles Metropolitan Transit Authority in 1996 committed to retrofitting 1,350 buses (half its fleet) with new catalytic mufflers, which is expected to eliminate 600,000 pounds of pollution a year—hydrocarbons by 50 percent, carbon monoxide by 40 percent, and total particulate emissions by 25 percent or more.

book, the Los Angeles experience illustrates the emergence of and the importance that the clean air policy had in the first epoch. It highlights the kinds of internal and external factors that led to the undermining of several key tenets of that epoch and the launching of the second. Los Angeles may represent an "extreme case analysis" in view of the depth of its air pollution problem, the extraordinary regulatory regime it developed in epoch one, and the important shifts under way in epoch two. Being such an extreme case, however, it servers to illuminate the outer boundaries of air pollution as a problem, and the lengths to which society has reached to remedy it.

The Los Angeles experience also underscores the uncertainties in our effort to understand the transitions between environmental epochs. It suggests that a concerted and relatively successful single-media effort (in this instance, to clean the air), while a logical and seemingly necessary precursor to a transition to the sustainable community of the third epoch is no guarantee that the needed transition will occur. The prospect of the third environmental epoch for Los Angeles is discussed at the end of the chapter.

Green Regulation Comes to Los Angeles

How best to clean the skies over Los Angeles has vexed regional leaders for more than fifty years, which made it a critical local issue long before air pollution emerged on the national scene. Air pollution policy took shape in the late 1940s, crystallized and expanded throughout the 1950s and 1960s, and since has remained center-stage. Also—which explains the issue's longevity—for all the reductions in air emissions from the early days up through Earth Day 1970 and the national awakening to the problem during the first environmental epoch and into the present, the air over Los Angeles remains the most polluted (categorized as "extreme" by the EPA) of any metropolitan region in the nation (see box 3.2).

Thriving industry and business, personal mobility, the internal combustion engine, a moderate climate, and location in a 100-square-mile natural basin surrounded by mountains all contribute to the attractiveness of Los Angeles, but also has made it the nation's air pollution capital.[1] The result can be everything from watery eyes, fatigue, respira-

Box 3.2
What is the nature of the air pollution problem in Los Angeles?

The answer is SMOG:
• It comes from burning fossil fuels in power plants.
• It comes from the heavy reliance on automobiles and the motorized vehicles for transportation.
• It comes from the fact that Los Angeles sits in a bowl, with mountain barriers to the north and east that trap the carbon monoxide, nitrogen oxides, sulfur dioxide, lead, and other toxic pollutants within the basin, often under a thermal inversion layer. This gaseous brown-grayish haze is carried steadily inland on the eastward blowing sea breeze, and is literally "cooked" by our glorious sunshine into photochemical smog.
• Ozone, the dominant summertime pollutant, has a pungent smell, but it is invisible.
• Nitrogen dioxide leaves a brownish smear on the horizon.
• Particulates, or microscopic bits of pollution composed of everything from diesel soot to dust in the air, are responsible for reducing visibility.
• "Smog" is a mixture of several pollutants, including ozone, particulates, nitrogen dioxide, and carbon monoxide.

The costs to human heath, the environment, and the economy are enormous:
• Epidemiologist David Abbey reported in 1991 a correlation between long-term exposure to air pollution and the development of chronic diseases.
• A study of over six thousand residents in the basin showed a higher incidence of respiratory diseases, including bronchitis and asthma.
• Women, in the most affected areas of the basin, showed one-third greater incidence of cancer.
• A 1989 district study concluded that if the region were in compliance with state and federal standards, it would gain $9.4 billion in health benefits alone.
• The EPA in 1995 concluded that particulate matter pollution as small as 2.5 microns in diameter and smaller (regulations were then set at 10 microns) were causing an estimated 275 deaths annually in the Inland Valley of the basin.
• A 1996 study by the Natural Resources Defense Council reports that fine particulate pollution kills 8,800 people prematurely each year in the South Coast Air Basin.
• In 1998, the results of a 15 year study of cancer among 6,338 nonsmoking Seventh Day Adventists found a link between particulate pollution, sulfur dioxide, and ozone and lung cancer.

tory disease, and headaches to emphysema, lung damage in children, and even cancer. Air pollution can reduce visibility appreciabily, require the curtailment of outdoor physical activity (particularly among the young and elderly), and cause extensive damage to physical property, plants, trees, and other forms of life. The cost in health care alone due to air pollution for the people of the region runs into the billions of dollars annually (Air Quality Management District [AQMD] 1998b).

From the 1940s, through 1960s, it remained the responsibility of local governments to identify targets for air emission control and address them. County officials in Los Angeles in 1945 took the lead by prohibiting factories from emitting dark smoke. This action was followed in the early 1950s by the banning of garbage burning, first in county dumps throughout Los Angeles County, then in everyone's backyard incinerator. Yet, as scientists began to discover, the problem was more than smoke from homes and factories. It was the meteorological conditions of a balmy climate and atmospheric inversions that trap in the air tons of tiny particles, lead, sulfur dioxide, carbon monoxide, nitrogen oxides, and ozone emitted by cars, trucks, buses, ships, planes, industrial smoke stacks, and the normal operations of business, which is then "cooked" by the sun into photochemical smog.

As it became clear that local initiative was insufficient, in the 1960s, the state of California stepped in and, despite strong opposition from the auto industry, began to place air pollution restrictions on new automobiles sold in the state. It also demanded more aggressive anti-air pollution steps be taken by local governments within the Los Angeles basin. As the issue grew in prominence, Congress increased the role of the federal government, most forcefully in the form of the 1970 amendments to the federal Clean Air Act. The cumulative impact on Los Angeles of the expansion in local, state, and federal air regulation over the next two decades would be an air pollution control regime larger in staff, better funded, and with greater reach into business, communities, and the lives of individual citizens than any other in the nation, or the world.

At the center of the pollution control regime is the South Coast Air Quality Management District (AQMD). AQMD was established in 1977 by the state legislature to manage in a single agency, with greater coordination and comprehensiveness than ever before, the quality of air in the 12,000 square miles of the four-county Los Angeles air basin. It succeeded the single-county Los Angeles County Air Pollution Control District that dated back to 1947. The AQMD is governed by a board of twelve, composed of four members appointed by the boards of supervisors of each of the four counties, five appointed by cities in the basin, and three appointed by the governor, the speaker of the California state Assembly, and the state Senate's Rules Committee. The mission of the AQMD is no less than to protect the public health from air pollution under the guidance of all relevant federal, state, and regional air pollution laws and standards.

Hundreds of air emissions controls have since been placed on business, industry, local communities, and government agencies, large and small, to reduce the level of the seven major "criteria" air pollutants set forth in the federal Clean Air Act—ozone, nitrogen dioxide, particulates, fine particulates, carbon monoxide, lead, and sulfur dioxide. In 1994, the regulatory net was extended to cover emissions of benzene, chromium, arsenic, and more than 100 other toxic chemicals, from approximately 250 of the largest emitters in the region. AQMD activities focus not only on stationary sources of the criteria pollutants but also on land use and transportation decisions throughout the basin. Its multifaceted approach has comprised everything from outright banning of highly toxic and polluting products to research support for less-polluting technologies for use in business and industry, and very low-emitting cars, trucks, and buses. It has pioneered in efforts to affect the decisions of individual citizens and businesses, such as getting people to use their cars less, and to modify their driving habits through ride sharing, telecommuting, and shifting driving to nonpeak hours. For example, AQMD's Commuter Program covers 6,000 companies employing 2 million people. Employers are required to encourage their workers to carpool, use public transit, or bicycle to work rather than drive alone.

AQMD has led in setting standards for air emissions in gasoline, solvents, paints, and many other commercial products, with the result that most have been reformulated by their manufacturers to be less polluting.[2] A permit system specifying emissions caps on industrial machinery and equipment has been put in place, as well as permits covering entire facilities. Today approximately 31,000 businesses are covered by

AQMD permitting, with their emissions limits enforced through periodic facility inspections, and with violators susceptible to fines and civil penalties. Educational campaigns have been conducted to raise the public's awareness of the need for the strong steps being taken to reduce air pollution. And new, cleaner technologies for industry, businesses, homes, and transportation are continually being promoted.

Today about 40 percent of the air pollution in the Los Angeles basin comes from the facilities and products that AQMD regulates directly. The remaining 60 percent comes from mobile sources, such as cars, trucks, trains, airplanes, and ships,[3] which are under the regulatory supervision of the EPA and the state of California's Air Resources Board (CARB). Even when the AQMD's reach to mobile sources is indirect, it is ever-present. For example, in full recognition of the severity of the air pollution problem (mostly, but not singularly in Los Angeles), new cars made for sale in California are required to be equipped with pollution controls that make them the cleanest internal combustion engine vehicles, burning the cleanest gasoline, in the world (Purdum 1998).

A Cup Half-Full or Half-Empty

Seemingly undeterred by the wide-ranging air pollution control measures that became synonymous with the first epoch of environmentalism, people continued to flock into the Los Angeles basin in ever-larger numbers, attracted by jobs and opportunities, the region's extensive natural amenities, and its growing reputation as the immigrant gateway to the United States, most notably from Mexico, Central America, South Korea, and Southeast Asian countries. From the 1940s through the late-1990s, the population of the four-county Los Angeles air basin has more than tripled from 4.8 million to nearly 16 million people, at a time the nation was doubling its size. Even more dramatic in its impact on air quality, the number of motor vehicles on the road quadrupled, from 2.3 million to nearly 10 million. Similarly, more and more trains, ships, and planes, businesses, and industrial activity have come, and with all this, an unremitting urban sprawl. The AQMD has not had the authority, the political will, or the administrative capacity to halt this growth, despite its obvious link to the air pollution problem of the basin. Consequently, the region's growth has significantly offset the gains in per person emissions reduction

Box 3.3
The fifty-year saga of air pollution in Los Angeles

- Fletcher E. Bowron, mayor of Los Angeles, announced at a press conference in August of 1943 that the city's smog would be eliminated within four years.
- In 1946, the county of Los Angeles established its first major air pollution control agency.
- In 1949, A.J. Haagen-Smit, a biochemist at the California Institute of Technology, reported that the automobile was a prime cause of smog.
- In the mid-1950s, the state of California created an agency to monitor and control motor vehicle emissions, then in 1959, gave the Department of Public Health authority to determine air quality and motor vehicle emission standards necessary to protect health, crops, and vegetation.
- In 1963, the adoption of the federal Clean Air Act gave the states technical assistance. It was amended in 1967, giving states primary responsibility for air emissions.
- In 1966, California established the first auto emissions standards, two years ahead of the federal government.
- In 1970, the seminal Clean Air Act was passed by Congress, giving responsibility to the U.S. Environmental Protection Agency to develop and implement national standards. A few of the more notable consequences were: introduction of catalytic converters in automobiles, the phasing out of leaded gasoline, and the raising of CAFE, or automobile fleet mileage standard to 27.5 miles per gallon.
- In California, in the 1970s, a virtually complete shift occurred from burning for energy coal and oil to far less-polluting natural gas in business, industry, and the electric utilities.
- In 1977, the South Coast Air Quality Management District was established by the state of California to address comprehensively the air pollution problem not only for Los Angeles but also for its surrounding areas. The district is governed by a twelve-member board, comprised of nine elected county supervisors and city council members from the four affected counties (Los Angeles, Orange, San Bernadino, Riverside), and three citizens, one appointed by the governor, one by the speaker of the state Assembly, and one by the Senate Rules Committee.
- After repeated delays in meeting air quality targets—first, 1975, then 1977, 1982, 1987, and 1988—and nonstop political haggling, Los Angeles has been granted until the year 2010 to meet federal air quality standards under the Clean Air Act of 1990.
- Today cars and light trucks account for 60 percent to 70 percent of air pollution over the Los Angeles basin. However, automobiles in 1993 were forty-three times cleaner running than their 1973 counterparts, or to put it differently, emissions from a car in 1993 were 10 percent of those produced in 1970.

achieved through all the efforts to reduce air pollution. The added pollution that accompanies the growth continues to account in large part for the basin's failure to meet federal and state air quality standards for ozone, carbon monoxide, and fine particulate matter, as well the need to more fully address toxic air emissions.

The Rise of AQMD

After more than a decade of straining to meet the emissions goals of the 1970 Clean Air Act, as well as the additional state and local requirements, resignation set in by the early 1980s among Los Angeles's government and business leaders. The issue came to a head over the AQMD's 1982 implementation plan to reduce ozone, a plan required by the Clean Air Act. Rather than spelling out how the region would come into compliance with the ozone standard in five years, as the Clean Air Act mandated, the plan did not foresee bringing the region's air quality into compliance for at least twenty more years. The EPA, which was supposed to ensure that all state and regional plans did meet the requirements of the law, went along with the region's assessment.

Still, the federal Clean Air Act granted only five years for coming into compliance. Consequently, in 1984, local environmentalists in the Coalition for Clean Air, a Santa Monica-based environmental group, supported by the Sierra Club, sued the EPA and the AQMD for the plan's failure to devise a persuasive and realistic strategy for meeting the law. The suit dragged on in court for four years until U.S. District Judge Harry Hupp eventually ruled in favor of the environmentalists and ordered the EPA to submit a new implementation plan for the region (Murphy 1988). This ruling became the legal backdrop for the renewed efforts and cleanup initiatives in the region, which would dominate the air pollution policy landscape for the decade to come.

At this juncture in AQMD's history, in 1986, Jim Lents was recruited to be executive director of the agency. Lents had been successful in helping Denver address its air pollution problems. His initiative and determination would eventually be credited with moving air pollution control forward in the Los Angeles region, and making it a force to be reckoned with ever since. Reflecting on this critical transition period, Lent mused, "The unstated position of the agency between 1977 and 1986 was 'we will try to improve the air quality, but Los Angeles is never going to have clean air.' . . . In 1987, however, we decided that would no longer be the case, that it is our mandate to have air in this basin that is healthy to breathe" (Waldman 1991, 170).

By the second half of the 1980s, air pollution policy became a driving force in Los Angeles in areas as diverse as energy and transportation decisions—including light rail, hydrogen powered locomotives, improved internal combustion engines, and especially the zero-emitting and low-emitting vehicles mandated for the region—urban planning, waste management, and landscape design. By the end the decade, the twelve-member AQMD board had emerged as almost a de facto regional governing body for the multicounty megalopolis of Los Angeles. A substantial administrative apparatus had grown up employing more than 1,000 staff with a $100-million plus annual operating budget, in a handsome, new, Diamond Bar headquarters, with enormous authority.[4]

With the most stringent air rules being applied to new businesses and industries entering the basin, combined with the provision in federal, state, and regional law that prohibited any significant deterioration in air quality, the AQMD became a key arbiter of what industries would be allowed to enter and which could afford to stay in the region. It even affected where they would be located. This was made vividly clear when the AQMD released its 1989 and then slightly modified 1991 regional air quality plan. That was the toughest, most intrusive set of regulations ever imagined in Los Angeles, or anywhere else for that matter. The regulations came in three tiers:

• Tier I comprised 130 control measures that AQMD believed could be adopted in the short term, using current technology and existing regulatory authority. They included measures such as ride sharing and alternative work hours, waste recycling, and using less-pollution-emitting building materials.

• Tier II controls were mostly extensions (but also a few more stringent applications) of Tier I controls. They included "on-the-horizon" technologies and policies that could reasonably be expected to be developed by the year 2000, such as telecommunications technologies and less-polluting alternative fuels.

• Tier III controls that would involve major technological breakthroughs of the sort that might emerge during the next two decades, including commercial applications of fuel cells, superconductors, and wide use of solar power.

Several benefit-cost analyses were conducted on the new plan but without satisfactory results. Their differences were partly technical but also political, involving who would bear the costs of the proposed controls. Also, while it is possible to gauge most of the near-term costs of imposing stricter controls, there was (and continues to be) uncertainty about the dollar value of human health, the quality of life, and the environment of the basin. According to the AQMD, the plan was to be carried out not only by the agency but also by the CARB, the EPA, and all local governments, transportation agencies, and other special districts and governmental entities within the region. The lists of control measures were extensive, and no level of government or agency was beyond reach of the plan (see box 3.4).

The new plan was not simply an incremental strategy that called for a few more end-of-the-pipe add-ons, or one that admonished Detroit to make a cleaner-running automobile, but a plan that called for tough new policies at home, with enactment and regulation by local governments, the state, and federal agencies. Indeed, the plan proved to be so sweeping in implication that it triggered an attempt by the chamber of commerce and business community, as well as a few local governments, to actually float the idea of creating a new regional government to counterbalance the extraordinary authority being amassed by the AQMD, along with the CARB and the EPA.

The Federal Implementation Plan (FIP)

If this were not enough, in early 1993 the EPA lost its appeal to the Supreme Court to overrule District Court Judge Hupp's 1988 decision against it and the AQMD in the case originally filed by the Coalition for Clean Air. Consequently, the EPA was forced to enter the region anew and devise its own mega-plan, the Federal Implementation Plan (FIP) for the Los Angeles region that would bring the basin into compliance with the federal Clean Air Act. The outline of the EPA's plan was released in February 1994, it reached to sectors only minimally touched up to that time by AQMD, including trains, ships, the Los Angeles and Long Beach port authorities, and commercial airlines. The draft plan went so far as to threaten the rationing of gasoline in order to curtail automotive emissions. Of course, the EPA knew that even using its full administrative and

Box 3.4
Examples of control measures in the AQMD 1989–1991 Air Quality Management Plan

---

Control measures to be implemented by local governments:
- Alternative work weeks and flextime
- Employer ride-share and transit incentives
- Local government energy conservation programs
- Emissions reductions from swimming-pool water heating
- Low-emission materials for building construction

Control measures to be implemented by the district:
- Controls for emissions from petroleum refinery flares
- Banning of new drive-through facilities
- Telecommunications controls
- General aviation vapor recovery

Control measures to be implemented by transportation agencies:
- Traffic flow improvements
- High-speed rails
- Diversion of port-related truck traffic to rail
- Freeway and highway capacity enhancement
- Control measures to be implemented by state and federal agencies
- Lower emissions standards on new jet aircraft engines
- Control of fugitive emissions from marine vessel tanks
- Control of emissions from pesticide applications
- Railroad electrification

---

regulatory powers it was not likely to bludgeon into compliance the local political and economic leaders, who were the main source of resistance. The EPA thus challenged the region—the AQMD, business, industry, and other community interests—to create its own new and comprehensive implementation strategy, and avoid having the EPA impose one on it. But this was to no avail.

The FIP was supposed to be in place by February 1995, but it arrived stillborn. The political resistance from truckers, shipping and airline industries, local business, the utilities, automotive manufacturers, the oil companies, local governments, and many, many more was overwhelming. The timing of the effort, in the midst of the longest recession the region had experienced since the Great Depression, significantly added to the opposition. If there was any question about this verdict in the months prior, the November 1994 election of an antienvironmental Republican Congress made it unequivocal. A deal was finally worked out in Washington whereby the state of California's already approved State Implementation Plan, itself quite demanding, would be substituted for the FIP, no matter how questionable its ability to bring the Los Angeles region into compliance within the time frame of the Clean Air Act. The environmentalists were told that if they did not go along with this compromise, and returned to the courts, Congress would revisit the Clean Air Act with the real possibility of reducing its stringency. The threat carried the day, and while the issue would eventually be appealed again in the courts (with the outcome uncertain at this writing), the tide had shifted.

This episode signaled a major political setback for the EPA and a turning-point for the command-and-control, top-down, regulatory regime in Los Angeles. The regulatory approach to air pollution control had peaked, running into insurmountable political and economic obstacles, forces that neither the courts nor the EPA were capable of overriding (Fiore 1995). A turning point had come not just for the region but also for the top-down regulatory model of achieving cleaner air for large, metropolitan regions everywhere. While signs of change had been percolating earlier, as will be discussed, this episode was a watershed for the first epoch of environmentalism.

Emissions Reduction Accomplishments of Green Regulation
When viewed in light of the region's enormous population growth and economic expansion, the record of emissions reduction in the Los Angeles basin from Earth Day through the 1990s was impressive. By the late-1990s, the aggregate level of several of the federal and state criteria pollutants had either been brought into compliance or had been substantially reduced. Regionwide trends for six criteria pollutants show that since the early 1970s the emissions of lead in the air declined to nearly zero. The levels of carbon monoxide and nitrogen dioxide emissions were significantly reduced. Ozone emissions had decreased, as well although not nearly as dramatically as the other pollutants. Peak concentrations of ozone, one of the toughest emissions to eradicate, declined by one-third from the early 1980s to the late 1990s.[5]

The reasons for the long downward trend are many, but credit is due mainly to advances in automotive tailpipe and smoke stack emissions control, the introduction of unleaded gasoline, and the introduction of cleaner-burning fuels in industry. As Jim Lents put it:

Today motor vehicles and industries operating in California are among the cleanest in the world. A new car sold in California emits just one tenth of the pollution that a new car did in 1970. Such industries as electric utilities rely almost exclusively on clean-burning natural gas. Manufacturing plants and construction companies use advanced paints, solvents and adhesives that have been formulated to minimize pollution. For these reasons and others, southern California has made tremendous progress in reducing air pollution. (Lents and Kelly 1993, 38)

All this was occurring at a time during which the region's population continued to spread even further across the landscape, north toward Ventura County, east into San Bernardino, Riverside, and the mountains and upper desert beyond, and south, filling in the remaining open space between Los Angeles and Orange counties. Therefore, while the decades of the 1980s and 1990s have seen great progress, the pressures from growth and sprawl suggest that much more is going to be needed if the region is to avoid reversing the downward trend of its emissions, as well as come into compliance with the federal and state standards for particulate matter, carbon monoxide, and especially, ozone. For example, during the smog season of 1998, the national health standard for ozone, while at its lowest point on record, was violated fifty-five times for one hour or longer every two to three days somewhere in the basin. Meanwhile, the number of "stage one episodes" reported at one or more monitoring stations in the basin jumped from an all-time low of a single episode in 1997, to twelve episodes in 1998. Moreover, for the first time, episodes bypassed the center of the metropolitan region, occurring along its eastern boundary in the small San Bernardino mountain communities. Today, ozone releases are still nearly three times the state and more than twice the federal acceptable level, and carbon monoxide and annual particulate figures both are well in excess of the called-for levels.

Studying these trends, two conclusions emerge. First, the extraordinary effort of the Green regulatory epoch brought the region a long ways along the path of attaining federal and state air emissions standards. Second, if one makes the assumption—as most do—that population growth will 'ontinue its historical march forward, increasing the region's population by as much as 25 percent by the year 2020 (along with the added economic development, cars, vehicle miles traveled, and congestion that such expansion implies) a real possibility exists that emissions, particularly of ozone precursors, could reverse their thirty-year decline.

At a minimum, if the region is to achieve state and federal emissions levels, it will need to reduce hydrocarbons emissions by upwards of 80 percent, nitrogen oxides by 70 percent, sulfur oxides by 60 percent, and particulates by 20 percent. Moreover, if Green regulatory strategies for emissions reduction have peaked, how is compliance going to be accomplished, especially given the economic and growth imperatives of the region?

The Pendulum Swings to Efficiency-Based Regulatory Reform and Implementation Flexibility

From the outset of the battle against air pollution, the political challenge of policymakers has been to achieve a balance between reductions in emissions unequivocally required by law and population and economic growth at the heart of the region's prosperity. To the extent that add-on emissions controls (e.g., catalytic converters in cars and scrubbers on smoke stacks), production-line modifications, and less-polluting products could be introduced without undermining the economy, the increasing regulation did not cause unacceptable disruptions. For all the complaints from the business community throughout the first epoch, the strategy worked. In the early 1990s, however, the reach and powers of the Green regulatory regime were severely shaken.

Los Angeles was engulfed by a recession starting in 1990, which forced all parties to pay attention to the costs of doing business, as well as the costs of government services. The political climate in the state had also undergone a dramatic change, becoming much more conservative in tone and Republican in composition. These two developments combined to make state and local elected officials extremely wary of adding new regulatory burdens, and in several instances to try to roll back some already in place. By this time, of course, almost all of the big, stationary sources of air pollution—businesses, industries, the utilities—as well as the major mobile source—the automobile—were already under the regulatory net. To substantially reduce emissions further, the net would have to be expanded to capture the thousands of small emitters and minute sources of air pollution. This was not politically feasible.

Finally, the AQMD's politically appointed board of governors turned over, mirroring the broader political trends in the state. The new group quickly tried to get out in front of the regulatory reform sentiment that was building in the state legislature, among locally elected officials, and the general public. This was evident in 1994, when the AQMD released its plan for emissions levels in the region. Market incentives, which received just passing mention in 1991, were the new plan's center-piece. Nearly one hundred programs were included. A new implementation plan was declared, aimed at 75 percent of all smog-causing activities: fees on car miles driven and on fuel consumption; credits and rebates for cleaner technologies; financial incentives to switch to electric or low-emitting vehicles and fuel-cell vehicles; expansion of RECLAIM (discussed later) to 1,200 industrial facilities; the planting of shade trees to reduce peak summer heat. Credits and rebates for energy efficient houses and businesses were proposed. Gone from the plan were controls on special events centers, shopping centers, airport ground access, trip reduction requirements for schools, a reclaim program for volatile organic compounds (VOCs), and other controls on commerce and business.

The new plan was an attempt to reconcile the need for further reduction in emissions—required by the 1990 Clean Air Act, EPA regulations, and the state of California—in the face of enormous political opposition to the stringent control measures previously under discussion, as well as in earlier plans. In essence, pushing the command-and-control, Green regulatory approach to the extreme was viewed as neither economically or environmentally practical, nor politically feasible in the Los Angeles region of the 1990s. The only strategies that were likely to work in view of the changed realities were ones that avoided forcing costly trade-offs between cleaner air and the growth in people, cars, and businesses. The region would have to find other than top-down, regulatory methods to meet its air quality goals.

Devising these alternative strategies now became the AQMD's focus. With its shift to "incentives" as opposed to "penalties," efficiency-based regulatory reform, and program implementation flexibility, the second environmental epoch had arrived in Los Angeles. The underlying thrust was proposed in the 1996 plan for the region, which while adding a number of new regulatory provisions for VOCs in particular, continued with overall regulatory relief by eliminating a series of categories that would have only modest impact on emissions generally. Moreover, the AQMD's board contended that these changes would not significantly alter the downward trend in emissions reductions in the region. While these projections of future trends in emissions were open to debate—and were in fact heavily debated with each successive AQMD plan—nine of the eleven of AQMD's scientific advisers could not bring themselves to support the assumption that market incentives and less top-down regulation would lead to the continued reductions suggested by the computer models that the board embraced. The scientists' doubts were heightened in light of the growing concern within the health community and the EPA about the need to regulate even more stringently very fine particulate matter. In August 1996, the nine advisers resigned (Cone 1996).

In practice, the transformation under way was less of a single strategy than many strategies guided by the principles of regulatory relief (in 1996 alone, permit requirements for 10,000 pieces of equipment were eliminated), efficiency, incentive-based programs, and an emphasis on new, less-polluting technologies. The transformation included streamlining the permitting process for business and industry to eliminate delays and redundancies among the different local, state, and federal air quality and other regulatory requirements. An extension of streamlining, often proposed but yet to be implemented, is to incorporate within one administrative entity regulatory responsibility for air, water, and other health and environmental pollutants, in a "multimedia" approach to environmental regulation. While several pilot multimedia projects have been tested by the EPA, none has proven fully satisfactory. It would appear that the drawback of this approach is less conceptual than practical, that is, the problems of bringing about coordination and cooperation among the different agencies and vested interests.

The most often discussed new strategy of the second epoch involves ample use of market incentives, cost-effectiveness and benefit-cost analysis, and risk analysis (Harrington, Walls, and McConnell 1994; COALESCE 1994; Sierra Research 1994; Southern California Association of Governments 1994; Cameron 1994; Johnson 1993). While these approaches have their critics, they appeared to have the most promise for long-term change. Moreover, Los Angeles has led the way by initiating the largest regional incentive-based approach in its RECLAIM program, which involves emissions trading for nitrogen oxide (NOx) and sulfur oxides (SOx). And a similar program for volatile organic compounds (VOCs) was under consideration for several years, though it was dropped from the 1996 plan and clearly has been set aside for the present.

RECLAIM: The Los Angeles Policy Experiment
In 1992, AQMD adopted the Regional Clean Air Incentives Market program, a novel policy whereby operators of industrial facilities are allowed to buy and sell excess or unused emissions permits—the right to emit a certain amount of a specific pollutant—in sulfur oxides (SOx) and nitrogen oxides (NOx). The purchasing facility can use the amount purchased to emit above its permitted allotment. The idea came straight out of elementary economics. That is, top-down regulations would be replaced by marketplace transactions to determine the monetary value of emissions permits. This would result in the highest and best use of such permits, in the most efficient manner, for both business and the region.

To achieve the regional goal of reducing the overall amount of emissions for both nitrogen and sulfur oxides, under RECLAIM each facility was first given an initial allotment (in pounds) of permitted emissions, based on its preexisting emissions level. The critical part of the program is that the allotment is reduced annually, by 5 percent to 8 percent, in order to achieve on overall reduction in the region of 83 percent in NOx and 65 percent in SOx emissions by 2003. The permit holder can use the permit to continue emitting up to its limit, or reduce its emissions and sell the excess to others. The program is estimated to reduce the cost of compliance by 40 percent to 50 percent, and, as an added incentive, participating companies are assured fast-track, one-stop shopping by the AQMD for all their major permit needs.

One virtue of this approach is that it is expected to lead to more rapid introduction of less-polluting technologies as companies discover that by installing less-polluting equipment they have excess emissions permits that they can sell at a profit. Also attractive to industry was RECLAIM's replacement of about sixty of the then existing and anticipated command-and-control regulations. The cost savings to industry overall have been estimated at upwards of $180 million, or 47 percent of compliance costs when compared with traditional forms of regulation.

Although it is too early to judge the efficacy of RECLAIM, one thing is clear. It is not only the first but has grown into the largest, market-based, regional emissions trading program in the nation. Since taking effect in 1994, it has been extended to 329 out of 30,000 industrial facilities in the Los Angeles region—including the 300 largest—and covers 53 percent of the nitrogen oxide emissions and 92 percent of sulfur oxide emissions from all facilities regulated by AQMD. The number of transactions among facilities has grown to 403, with nearly 29,000 tons of NOx credits valued at $24 million, and 13,000 tons of SOx credits valued at more than $13 million changing hands (AQMD Advisor 1998).

Still, a number of concerns regarding RECLAIM's design and implementation have been raised, prompting a three-year audit by AQMD, which began in 1997. Among the potential problems that participants and observers have identified, the two most frequently voiced are the absence of short-term environmental benefit and the disincentives in the rules for more active trading. While many environmentalists supported the RECLAIM concept, several opposed the final program design because the short-term environmental impact would probably be negligible—and they have not been persuaded otherwise. Their attitude resulted from the political compromise that led to the initial allocation of credits, which was based on a firm's average maximum emissions, for the 1991–1994 period, at a time when the economy was in recession and overall industrial production was down. Others have argued that the program's size (limited to 300 plus firms) would keep demand and supply of credits limited, and thereby inhibit the development of an efficient market. On the supply side, trading would be limited because credits from shutdown facilities would be discounted by 80 percent. Supply would also be limited because firms might hoard credits for future use (fearing that credits then would either be not available or too expensive), or as a safeguard against potential future reductions in the overall level of emissions. Finally, critics emphasize that the vast majority of the annual cost savings would go only to three industries, and therefore provide little incentives for the other businesses to engage in trading. The concern from business and industry in general, on the other hand, was that the system remained plagued by too many rules and that the detailed reporting and monitoring requirements were no different than earlier command-and-control regulations. As one analyst concluded: "So far California's emissions trading program is a relatively narrow scheme crafted onto a comprehensive command-and-control program" (Dwyer 1992, 7).

Although these are important concerns, the significance of the movement toward regulatory relief and market-based approaches embodied in RECLAIM goes well beyond its promised greater economic efficiency. It signifies a shift in basic principles about how to bring about cleaner air. A second dimension of the new approach involves moving to the front of the technology curve, rather than trying to squeeze the last ounce of emissions reduction out of existing technologies and industries. Where further squeezing is nevertheless needed, the question has shifted to how to provide the right, positive incentives to induce change at the level of the consumer and business enterprise rather than having AQMD, the state, or federal government prescribe how to do so (Southern California Economic Partnership 1994).

This fundamental change in philosophy is taking place in the face of the elaborate administrative apparatus of emissions control painstakingly erected over several decades. It appears to be gaining momentum even before resolution is reached on the very significant new strictures of federal and state air pollution control laws bearing down on the region, as well as the detailed plans that have been drawn up by the AQMD, the state, and the federal government to bring the region into compliance. What is most telling, however, is the importance being given to technology and market-based approaches in accomplishing major emissions reduction.

Additional Incentive-Based Approaches for Los Angeles

Looking to the future, a number of incentive-based approaches (IBAs) have been considered for the region within and beyond the air quality arena and are worth noting here. While few have been adopted, they are just waiting for the right moment to move onto the policy stage.[6] What is important is that they all take advantage of market-based incentives to promote environmentally desirable behavior, thus reducing the necessity of and dependence on command-and-control regulation.

To begin with, AQMD at one point hoped to expand RECLAIM to cover many more than the present set of firms in the program. It also spent several years developing a RECLAIM for volatile organic compounds (VOCs), although it was set aside due to the inability to work through the program's operational mechanisms with industry. This was being developed at the same moment in the early 1990s when the political winds were shifting and little consensus could be reached on any new, major AQMD initiative, market-based or otherwise.

IBAs have been under discussion for several years for mobile sources of emissions in the transportation policy arena, beyond the confines of AQMD, but with significant implications for air emissions reduction. Not only are mobile source emissions the most significant cause of pollution in the Los Angeles basin, but federal and state policy now requires transportation planners to design systems that simultaneously reduce highway congestion and air pollution. Both goals can be achieved, it is believed, through the introduction of innovative pricing schemes, such as congestion pricing, a "vehicle-miles-traveled" (VMT) fee, smog fees, higher gasoline taxes, and incentives for the purchase of zero-emitting vehicles (Johnson 1993).

The approach that received the greatest attention in Los Angeles in the early 1990s among business and environmental leaders, the regional association of governments, and academic analysts, who saw it as one of the most promising, was the VMT or vehicle-miles-traveled fee, which every driver in the state would pay. The VMT fee is direct and equitable: every one pays the same amount per mile of driving. The more one drives, the more one pays for roadways "used and consumed." Likewise, savings to the driver accrue with every mile not driven, which is the "incentive" to accomplish the desired dual-policy goals of reduced driving and reduced pollution. A number of methods exist for reading a car's mileage electronically and the system can be designed to be administered easily and unobtrusively.

The VMT fee has been promoted, also, as a replacement for the state's gasoline tax. Thus, operating revenue for the state's highway and road system could be shifted away from gallons of gasoline consumed to miles traveled. To the extent that the VMT provides incentives to move to public transportation, ride-sharing, and reduced trips, and to encourage people to be more efficient in driving overall, the less the roadways are congested. Air quality in Los Angeles would benefit doubly under this scheme, from both the fewer total miles of driving in the basin and the reduction in congestion, which itself is a major cause of auto emissions.

Several researchers consider VMT fees the most cost-effective, least disruptive, and politically palatable of all the approaches that have been scrutinized in recent years (Cameron 1994; Harrington 1994; OECD 1993). The level of the fee is always a major point of consideration, of course. Leaders in the Los Angeles business community in the early 1990s proposed a revenue-neutral penny a mile, or $100 for every 10,000 miles of driving (COALESCE 1994). Small and Kazimi (1994), using a health-cost criterion to establish the appropriate fee level, estimated that the full cost of adverse health effects from auto emissions could be captured with a 3-cents per mile VMT fee on cars (equal to approximately 69 cents per gallon of gas). But cars are only part of the problem, and they proposed that a far more substantial 53 cents a mile fee would be needed for large diesel trucks (in recognition of their disproportionate contribution to the air emissions in the region). The Environmental Defense Fund, an environmental policy organization, argued that a 5-cents per mile fee would be needed to achieve the optimal amount of reduced congestion and pollution (Cameron 1994).

The approach that has actually received the most public attention in the region, although far more problematic, has been the introduction of electric vehicles (EVs). EVs are the only automobiles that can satisfy the "zero emissions" rule of the California Air Resources Board. While in 1996 CARB backed away from its requirement that the major auto companies sell up to 40,000 zero-emitting vehicles per year starting in 1997 (the original rule set the goal at 2 percent of the cars sold annually), CARB has retained the 5 percent requirement for 2003, and double that by 2010. EVs engines release no direct air emissions, have the potential for home or away refueling, have an easy-to-maintain mechanical system, and can be designed with almost all the comforts of the conventional internal combustion engine automobile. These qualities make them extremely desirable from an air emissions and driving standpoint.

The shortcomings of the EV strategy are several, however. First, the battery technology for EVs remains cumbersome and relatively limited in range (from 50 miles to 120 miles), even with continuing battery improvements. Second, and more important, the purchase price of an EV today is as much as $10,000 to $15,000 higher than the comparable conventional automobile, even higher for light and medium-sized vans. This is the primary reason why General Motors and other automakers have chosen to lease rather than sell their EVs. In terms of emissions, it has been estimated that reducing 1 ton of VOCs through introducing EVs may turn out to cost from $29,000 to $108,000 (Harrington, Walls, and McConnell 1994, table 1). These considerations raise a host of policy issues with respect to equity and the overall efficacy of the EV as a cost-effective approach (which accounts for the low rating it is given in this category in table 3.1). While modest public subsidies for purchase of EVs are available from the federal and state governments, the questions of who will bear the burden of the cost differential—the general tax payers, buyers of conventional cars, EV purchasers, or the auto manufacturers—has yet to be resolved. Even if it is resolved, the cost-effectiveness of the approach will change little. That will happen only when EVs become cost-competitive.

The VMT fee and EVs are only two examples of the IBAs given attention by analysts and interested industrial and environmental stakeholders as the region entered the second environmental epoch. Table 3.1 lists eight approaches,[7] and evaluates each briefly, based on their emissions reduction potential for the Los Angeles basin, the directness and simplicity of their implementation, their cost-effectiveness,[8] the considerations of equity and justice that they raise, their potential impact on the regional economy, the extent of their intrusiveness into the lives of businesses and individuals,[9] and their level of political (existing and potential) support. Included are VMT; an emission tax and rebate for both new and old cars; old car scrappage; purchase of electric cars; an enhanced RECLAIM for SOx and NOx; a RECLAIM for VOCs; and pay-at-the-pump auto insurance for all California drivers. Given the fluidity of issues in the air and transportation policy arenas, the ratings in the table are only suggestive, serving as summary judgments based on interviews with a wide range of analysts and people involved in the discussions.

As the ratings show, there is no single, best approach that satisfies all concerns simultaneously—from extent of emissions reduction, simplicity, cost-effectiveness, and equity to economic impact on the region, intrusiveness, political support, and proportion of the problem addressed. Each approach involves trade-offs between the directness of the economic "signal" to the level of reduced emissions, political acceptability, and so forth. Given the amount of publicity and political support behind the EV initiative in the Los Angeles region, it is noteworthy that it does not appear to be a highly cost-effective approach when judged by the criterion of air pollution reduction, at least at this stage in its development. Most of the other approaches could provide a great deal of emissions reduction, cost-effectively, if they were implemented. If all were adopted, they would have a tremendous impact on air emissions reduction, and make IBAs the dominant form of air pollution control in the region. Of course, this is a very big "if."

Toward Sustainable Communities: Living in the Present While Reaching the Future

The adoption of market-based approaches in Los Angeles, as important as they have become, do not appear likely to produce the remaining and in some instances steep air emissions reductions needed to meet federal and state clean air standards. More dramatic steps will likely be required. One answer to that problem is to return to the command-and-control strategy of epoch one and simply force people to change through government fiat, but this seems an unlikely choice. An alternative is to move Los Angeles in the direction of the broader environmental movement that is sweeping society today, toward greater sustainability and the third environmental epoch. This, of course, will pose its own set of challenges. It will require a significant transformation in what people value, where

Table 3.1
Incentive-based approaches for air emission reduction in Los Angeles

	Air emissions reduction	Directness and simplicity of implementation	Cost effectiveness	Equity and justice considerations
1. Vehicle-miles-traveled (VMT) fee	high—at the 5 cent level, or when combined with emissions fees	high—(presuming tamperproof odometers)	high—with respect to reducing congestion and emissions	mixed—depends on driving behavior and transporation alternatives
2. New car emissions fee and rebate	moderate—existing standards are substantial	high	moderate	neutral
3. Old car emissions fee and rebate	high—if based on age, model, and past performance	high	high	mixed—regressive to the extent it hits the poor
4. Old car scrappage	mixed—depends on design of program	high/mixed	high—if gross polluters are the target	mixed—regressive to the extent it hits the poor
5. Electric car purchase	high—zero emissions	high/mixed	low—if $10,000 differential in purchase price	regressive—without a subsidy; neutral—with a subsidy
6. Enhanced RECLAIM	moderate	mixed—green tape problems	moderate	neutral
7. VOC RECLAIM	moderate	mixed—unclear at this time	high (if designed properly)	neutral

	Impact on regional economy	Extent of intrusiveness	Level of political support	Proportion of overall air pollution problem affected
8. Pay-at-the pump auto insurance	moderate to high—at full cost of insurance (25 to 50 cents per gallon)	high	very high—no added costs needed to achieve emissions reduction	regressive—to the extent the poor are required to pay insurance
1. Vehicle miles-traveled (VMT) fee	mixed—depends on adjustments in work hours, etc.	low—presuming automatic recording system	moderate overall, though high when compared with the alternatives	high
2. New car emissions fee and rebate	neutral	low	mixed—opposed by auto manufacturers and big car buyers	low
3. Old car emissions fee and rebate	neutral	low	low—concern with regressivity	high
4. Old car scrappage	minimal	minimal	high	moderate to low
5. Electric car purchase	mixed—high if deep market penetration; low without	moderate	mixed—low if public subsidy required; high without	mixed—high only if deep market penetration
6. Enhanced RECLAIM	moderately positive	moderate	moderate	minimal to moderate
7. VOC RECLAIM	moderate	moderate	mixed—much uncertainty remains	moderate
8. Pay-at-the-pump auto insurance	moderate	low	mixed	moderate

they live and how they work, and how the economy is organized. It would likely require appreciable changes in the region's governance.

The path toward sustainability as outlined conceptually in chapters 1 and 2, to be effective as a blueprint for an environmental strategy for Los Angeles, suggests several necessary, though by no means fully sufficient steps, including:

• Treating the entire Los Angeles basin as a single air shed—that is, applying a bubble concept to the region as a whole.
• Enabling emissions trading across multiple pollutants, on a regionwide basis (a "universal emission trading credit" system).
• Establishing an emissions bank to facilitate trading in emission credits and, in turn, to direct the funds collected to the most cost-effective emission reduction targets.
• Ensuring that all businesses in the region adopt "total quality management" and "life-cycle cost accounting" approaches as standard operating procedure.
• Encouraging the redesign of many if not most of today's products and technologies to eliminate the use of toxic materials, generate fewer wastes, and consume less energy and other renewable and nonrenewable materials.
• Facilitating the transformation of public agencies overseeing pollution abatement (air, water, toxics, and others) from their traditional regulatory style and culture to becoming collaborators and facilitators in the emerging high-tech, market-based economy of the region.
• Developing ways of reducing inequalities in education, wealth, and quality of life for all residents of the region.

For such a sweeping transformation to occur will require more than reengineering environmental policies and governmental agencies. It will necessitate moving to the forefront of the technology curve to eliminate toxic and polluting air emissions. One step in this direction would be the redesign and remanufacture of the personal automobile. A radical thought, admittedly, but it comes from none other than Amory Lovins, the sage of the energy revolution of the 1970s (who erred only in underestimating the rapidity with which energy conservation and soft paths would be realized). Lovins sees today's internal combustion, heavy, aerodynamically awkward personal automobile as extremely antiquated. Not only the car but also how and where it is manufactured is passé, and this has broad implications for energy policy and environmental protection, not to mention passenger comfort, safety, and the fate of the U.S. economy. His thesis is powerful and cannot easily be dismissed:

Conventional cars, like other technologies, have entered their era of greatest refinement just as they may have become obsolete.

Imagine that a seventh of the GNP in, say, the United States were devoted to manufacturing typewriters. The Big Three typewriter manufacturers have gradually moved from manual to electric to typeball models. Now they are making delicate little refinements somewhere between a Selectric 16 and a Selectric 17. Their typewriters are excellent and even profitable. People buy over ten million of them every year. The only trouble is that the competition is working on subnotebook computers.

That, we suggest, is where the global auto industry is today—painstakingly refining designs that may soon be swept away, perhaps with terrifying speed, by the integration of very different technologies already in or entering the market, notably in advanced materials, software, motors, microelectronics, power electronics, electric storage devices, and computer-aided design and manufacturing. (Lovins, Barnett, and Lovins 1993, 3)

This transformation is being given serious attention today among industry experts (DeCicco and Ross 1994). Should it become a reality within a decade, there is little doubt that it will dramatically alter the need for the AQMD, CARB, and EPA to develop extensive new programs for emissions control.

For all the potential of a supercar "silver bullet," or for that matter, the lightweight, inexpensive, and easily rechargeable electric vehicle battery, for solving LA's air quality problems, near-term pubic policy cannot depend upon yet-to-be-proven technologies. The focus of policy makers must be on the near-term gains that can be achieved through the wider application of IBAs. This will require that IBAs meet the dual test of short-term efficacy and cost-effectiveness, yet not create a system so fixed, or necessitating such extensive long-term capital investment, that it precludes the possibility of future technologies readily entering the region.

The prognosis for the long term—ten, fifteen, twenty years out—would seem to require additional fundamental change as well: actions designed to move beyond minimizing to eliminating harmful emissions. A sustainable environment, if it is to be realized, will need to move beyond end-of-the-tailpipe solutions. It will require accelerating the introduction a hydrogen fuels-based economy. It will require increasing efficiency in manufacturing, transportation, and business services.

Table 3.2
A comparison of the first and second environmental epochs in Los Angeles air pollution control policy

	Regulating for environmental protection, 1970–1990	Efficiency-based regulatory reform and flexibility, 1980–1990s
Problem/ Objectives	To reduce air emissions from existing industry and transportation through establishing national health-based standards, imposing end-of-tailpipe emissions restrictions, and using air pollution policy to compel changes in work and lifestyles	To devise a regional growth strategy that utilizes market incentives to encourage development and application of minimally polluting technologies; to foster green business networks and public-private collaboration
Philosophy (orientation) of business	Considers pollution a public- (not private-) sector issue; rejects the basic premise and methods of the environmental movement; pollution viewed as an externalities problem; adversarial in tone	Recognizes the need to reduce pollution as a matter of law and good business; looks to public policy to provide incentives to reduce pollution and help overcome problems of collective action in industry
Philosophy (orientation) of government	To establish environmental standards and oversee business and citizen compliance through top-down, command-and-control regulation	To be a partner with business in the greening process; to devise win-win policy incentives for economic growth and environmental protection
Philosophy (orientation) of environmentalists/ health interests	To reverse environmental harms caused by modern industrial society and an expanding global population: zero emissions, polluter-pays principle, refusal to acknowledge costs, reliance on federal command-and-control policy and courts, grassroots social movement politics	Same basic ends although deeply divided on means along the lines of cooperation/co-optation versus adversarial/deep ecology orientations and strategies

Predominant national political/ institutional context	USCAA, NEPA, EIRs	ISTEA, EPACT, Clean Cities, DOD's ARPA programs, some sections of the USCAA
Predominant LA regional political/ institutional context	CaCAA, CEQA, AQMPs, regional rail intitiatives and MTAs, utility DSM programs, alternative energy programs	Project CA, LA's New Economy Project, CALSTART, SCE Team Cities, SC Economic Partnership, Aerospace & Defense conversion
Political style	Adversarial, winners and losers; insiders and business dominate the process	Collaborative and inclusive of all interests (corporatist); a search for common ground
Institutional venue	The legislature and executives offices in Sacramento and Washington, ARB and the Air Districts; the conventional public policy making arenas; partisan politics	New public-private partnerships and private collaborative forums; more fluid political coalitions; vertical and horizontal linkages across groups and federal, state, and local venues
Power relationships and key players	Elected officials, the recognized representatives of environmental and business groups, industry lobbyists from the leading petrochemical, energy, and manufacturing sectors	Traditional business and industry leaders plus those from the new high-technology industries, a wider array of social and community leaders, universities, technical community
Duration	1970–1995—From Earth Day, passage of NEPA and the 1970 CAA to the issuance of the draft, regional federal implementation plan for the Los Angeles region from the EPA in 1994	1990–2010 (and beyond)—From the rise of the new public philosophy and the region's economic restructuring, RECLAIM, the shrinking of SCAQMD, revision of Rule XV, to the market-and technology-based stategies of the twenty-first century

For both technical and economic reasons, the transformation will propel firms forward on the ladder of complexity and cost. This begins with the good, internal housekeeping practice of managing resources and waste, then moves to better inventory control and adjustments in minor operating practices (Smart 1992; Dorfman, Muir, and Miller 1992). Progress will require "significant technological advances in reaction engineering, separation science, and process synthesis" (Allen 1992, 1159). It will entail a host of new, cradle-to-grave, "total cost assessment" and "life-cycle analysis" management, accounting, and decision-making systems, and "life cycle design" for products, all of which incorporate direct and indirect environmental costs. Managers will need to be able to select the overall and multimedia (air, water, noise) least harmful and least costly mix of products and production processes (Freeman et al. 1992; Keoleian and Menerey 1994).

Achieving such a future for Los Angeles will surely require a profound change in the philosophy of government in the region, with greater emphasis on the virtues of decentralization, dynamic policy strategies, and collaboration and coordination across communities, agencies, and interests. Green tape will have to replace red tape. While what is needed for the transformation is fairly evident, how it will come about—short of a major environmental catastrophe—in not easy to imagine. Experiments at the grassroots in Los Angeles and elsewhere are under way already (see chapter 2), but whether they are the tip of the iceberg or exception to the rule is unclear. Whatever strategies the environmental movement chooses to persuade society to embrace sustainability, environmentalists will need to foster much greater awareness of the environmental challenges facing society and an appreciation of the possible mechanisms of change.

Conclusion

The Los Angeles region, which experienced the most stringent, comprehensive, and powerful air pollution control program of the first epoch of the environmental movement, finds itself in a major transition into the second epoch. The "problem" of air pollution is being redefined from one of dirty industry and automobiles to finding clean technologies and bringing them on-line through market incentives. The players in the air emissions arena have moved from staunch antagonists to sometime allies and collaborators in problem solving. Government is no longer seen as the ultimate policy enforcer—through command-and-control regulation—but one party, albeit important, in the search for community-based and cost-effective solutions. Taken together, these shifts in thinking, politics, and policy, if not always in actual behavior, have moved Los Angeles far into the efficiency-based, regulatory reform, and flexibility epoch of environmentalism.

Table 3.2 captures this transition by summarizing the features for the first two epochs of environmentalism: the initial epoch of green regulation, then the 1980s to 1990s epoch of regulatory reform, market-, and incentive-based approaches and technology strategies. Consistent with the general thesis of chapter 1, the broad policy goal of cleaner air remained constant, but the manner in which it was to be achieved, and at what cost, as well as through which methods and mechanisms, differed substantially.

It remains to be seen how far Los Angeles, in so many ways America's preeminent, trend-setting city of the twentieth century, will be able to remake itself into a sustainable city of the twenty-first century. Not only will it need to address the issues of air pollution but it will also have to look at water, transportation, business and industry growth, urban sprawl, population expansion, and the education and quality-of-life needs of its citizens.

Even before the transition from the first to the second environmental epoch is completed, the third beckons. But as will be evident in subsequent chapters, much more is needed than a media-specific—in this case, air pollution effort.

Notes

1. The attractiveness of Los Angeles as a place to live has a long history just as does the region's history of air pollution, as revealed in the impressions of the earliest European explorers. For example, "When the Spanish explorer Juan Rodriguez Cabrillo sailed in 1542 past the coastal plain of what became Los Angeles, he was so struck by the dark haze from the Gabrieleno Indian campfires hovering over the mountain-ringed basin that he christened it the Bay of Smoke" (Purdum 1998, A-10).

2. The state of California has also entered the products arena with a major program for addressing product air pollution, as of July 1997. The program set emissions standards for more than 3,000 commercial and household products that cumulatively have appreciable effect on Los Angeles air—including automotive polishes, heavy-duty hand soaps, paint removers, metal polishes, spot removers, herbicides, lubricants, floor wax strippers and hair spray (Cone 1997).

3. For example, the AQMD reported in 1998 that marine vessels—oceangoing ships, harbor tugboats, and commercial boats—emit twice as many smog-forming emissions as do all of the region's power plants combined (AQMD 1998a).

4. The growth in staff and activities in the first epoch was to be matched in the second epoch (discussed later) by cuts in staffing between 1991 and 1998, of 36 percent, along with a 17 percent reduction in operating budget.

5. AQMD provides extensive quantitative data on the region's pollution levels by criteria pollutant and six categories of health standards, annually as far back as 1976; see the AQMD website at http://AQMD.gov/smog/html.

6. How long their wait will be, or if they will ever move onto the policy center stage is up to question, however. As John Kingdon persuasively argues, this requires the convergence of three processes that cannot be easily managed or predicted. They include the wide recognition that a problem exists, the generation of policy proposals to solve it (such as those being described in the case of air pollution), and the occurrence of a triggering event, what is often called "timing" in the political arena (Kingdon 1984).

7. Proposals are quite numerous; those included are only the most prominent today. Incentive programs for telecommuting and teleshopping could be justified, as well as incentives for reducing energy use, thus air emissions, through designing more efficient buildings and urban landscapes.

8. Cost-effectiveness is defined in conventional terms as the highest possible reduction in emissions for every dollar spent. This is not the equimarginal principle often used by economists when they speak of cost-effectiveness, where all emitters attain the same level of reduction on the marginal cost curve.

9. Intrusiveness refers to the extent to which rules and regulations require individuals and businesses to undertake activities or changes in behavior in a manner that makes them resent the agent (typically a governmental entity) responsible for implementing and enforcing a public policy. While often the cost of compliance is related to the level of resentment, just as often it is not, especially among individual citizens.

References

AQMD (South Coast Air Quality Management District).

AQMD. 1998a. "Marine Vessels a Major Source of Air Pollution." Http:/www.AQMD.gov/news/tugboat.html, August 7.

AQMD. 1998b. "Health Effects of Southland Smog." Http:/www.AQMD.gov/ smog/inhealth.html, October 4.

AQMD Advisor. 1998. "First Comprehensive Audit Finds RECLAIM Program Working," July, 3.

Allen, David. 1992. "Industrial Pollution Prevention: Critical Review Discussion Paper." *Journal of Air and Waste Management Association* 42l: 1159–1961.

Bryner, Gary C. 1993. *Blue Skies, Green Politics: The Clean Air Act of 1990.* Washington, DC: CQ Press.

Cameron, Michael. 1994. "Efficiency and Fairness on the Road: Strategies for Unsnarling Traffic in Southern California." Oakland, CA: Environmental Defense Fund.

COALESCE (Coalition for Local Environmental Solutions and a Competitive Economy). 1994. "Seeking Clean Air, Mobility and Quality Jobs for California's Future," a proposal (August).

Cone, Marla. 1997. "Smog Fighters Zero in on Home Products." *Los Angeles Times,* July 27, A-22.

Cone, Marla. 1996. "9 AQMD Advisors Quit in Protest of New Smog Plan." *Los Angeles Times,* August 9, A-1.

Davies, J. Clarence and Jan Mazurek. 1998. *Pollution Control in the United States: Evaluating the System.* Washington, DC: Resources for the Future.

DeCicco, John, and Marc Ross. 1994. "Improving Automotive Efficiency." *Scientific American* 27: 52–57.

Dorfman, Mark, Warren Muir, and Catherine Miller. 1992. "Environmental Dividends: Cutting More Chemical Wastes." A Report by INFORM, New York, NY.

Dwyer, John P. 1993. "The Use of Market Incentives in Controlling Air Pollution: California's Marketable Permits Program." *Ecology Law Quarterly* 20: 103–117.

Dwyer, John P. 1992. "A Free Market in Tradable Emissions is Slow Growing." *Public Affairs Report.* Institute of Governmental Studies. University of California at Berkeley, January, 1 and 6.

Fiore, Faye. 1995. "EPA Bends on State Anti-Smog Plan. *Los Angeles Times,* February 3, A-3.

Freeman, Harry, Trees Harden, Johnny Springier, Paul Random, Mary Ann Currant, and Kenneth Stone. 1992. "Industrial Pollution Prevention: A Critical Review." *Journal of Air and Waste Management Association* 42: 618–656.

Harrington, Winston, Margaret Walls, and Virginia McConnell. 1994. "Shifting Gears: New Directions for Cars and Clean Air." Discussion Paper 94–26. Washington, DC: Resources for the Future.

Kamieniecki, Sheldon, and Michael R. Ferrall. 1991. "Intergovernmental Relations and Clean-Air Policy in Southern California." *Publius: Journal of Federalism* 21: 143–154.

Keoleian, Gregory, and Dan Menerey. 1994. "Sustainable Development by Design: Review of Life Cycle Design and Related Approaches." *Journal of Air and Waste Management Association* 44: 645–668.

Kingdon, John W. 1984. *Agendas, Alternatives, and Public Policy.* Boston: Little, Brown.

Johnson, Elmer. 1993. "Avoiding the Collision of Cities and Cars: Urban Transit Policy for the Twenty-first Century." American Academy of Arts and Sciences, September.

Lents, James, and William Kelly. 1993. "Clearing the Air in Los Angeles." *Scientific American* 269: 32–39.

Lovins, Amory B, John W. Barnett, and L. Hunter Lovins. March 1993. "Supercars: The Coming Light-Vehicle Revolution," Rocky Mountain Institute.

Murphy, Kim. 1988. "Court Orders EPA to Develop Plan to Clean Air Basin." *Los Angeles Times.* September 20.

OECD (Organization for Economic Co-Operation and Development). 1993. *What Price Clean Air? A Market Approach to Energy and Environmental Policy.* New York.

Purdum, Todd S. 1998. "Spring Is in the Los Angeles Air but the Smog, Chased by El Nino, Is Not." *New York Times,* April 21, A-10.

Sierra Research. 1994. "The Cost-Effectiveness of Further Regulating Mobile Source Emissions." Report No. SR94-02-04. Sacramento, CA, February.

Small, Kenneth, and Camilla Kazimi. 1994. "On the Cost of Air Pollution from Motor Vehicles." University of California Transportation Center, Irvine, September.

Smart, Bruce. 1992. *Beyond Compliance: A New Industry View of the Environment.* World Resources Institute.

Southern California Association of Governments. 1994. "Transportation Control and Indirect Source Measures Recommendations from the SCAG Regional Council." Appendix IV-C, South Coast Air Quality Management Plan—1994, April.

Southern California Economic Partnership (The Partnership). 1994. "Organizational Summary and Marketing Strategy," October.

Waldman, Tom. 1991. "LA Air Board Starts a Fresh Wind Blowing." *California Journal,* April.

4

Clean Water and the Promise of Collaborative Decision Making: The Case of the Fox-Wolf Basin in Wisconsin

Michael E. Kraft and Bruce N. Johnson

The enactment of the Federal Water Pollution Control Act Amendments of 1972, known as the Clean Water Act (CWA), was a key signpost of the new era of environmental policy launched in the 1970s. Passage of the act signaled the nation's determination to clean up its heavily polluted streams, rivers, and lakes through a federally driven regulatory process based on national water quality standards and joint federal-state implementation. The new law provided a comprehensive framework of pollution control standards, tools, and financial assistance. It also established deadlines for eliminating the discharge of pollutants, and it ambitiously sought to "restore and maintain the chemical, physical, and biological integrity" of the nation's waters.

In most important respects, the CWA's reliance on a command-and-control approach that is applied to a single environmental medium (water) places it squarely within what the introductory chapter describes as epoch one. This strategy concentrated on regulating for environmental protection with a focus on business, industry, and municipal waste dischargers, and attempted to control their all too obvious end-of-the-pipe pollution.

The 1972 act was a dramatic break with previous policies that were widely faulted as ineffective and excessively dependent on highly variable state economic resources, bureaucratic capacity, and commitment to water quality goals (Ringquist 1993; Kraft 1996). The new policy was also successful by most measures. It resulted in significant improvements in overall national water quality. Yet the CWA was costly, too, with a cumulative national expenditure of public and private funds estimated at $500 billion over its first twenty years alone (Knopman and Smith 1993).

Not surprisingly, as is the case with other national environmental policies of the 1970s, both formal and informal assessments of the CWA over the past two decades have questioned its efficiency and effectiveness. Particularly in the 1990s, such critical evaluations have called for the use of new approaches that promise to reduce costs of compliance, promote greater flexibility and efficiency in implementation, and emphasize pollution prevention and ecosystem management, especially within river basins (Freeman 1990; Davies and Mazurek 1998). These strategies are hallmarks of epoch two, as discussed in chapter 1. A key component of the new approaches, which is thought to be essential to long-term success, is the substitution of a more cooperative and collaborative process of decision making for the often adversarial relations between government and polluters, characteristic of conventional environmental regulation. Yet confidence in the efficacy of such collaboration would be enhanced if its use in specific cases were evaluated.

To provide such information, this chapter examines experience with water quality policy in one region of the United States that shares many qualities with other industrialized areas of the nation. We review the recent history of the struggle to clean up the Fox-Wolf River Basin in Northeastern Wisconsin, which surrounds metropolitan Green Bay. We highlight the strategies used nationwide as well as in this region during the first two epochs, and discuss the emergence of the third epoch of sustainability. We also focus on major achievements to date in cleaning up area waters and the remaining tasks, especially the enormous challenge of removing highly contaminated sediments laden with polychlorinated biphenyls (PCBs) deposited years ago through discharges by area pulp and paper mills.

Over time, the Wisconsin Department of Natural Resources (WDNR) and other governmental bodies and industry have made use of all of the techniques available under the CWA and commonly used throughout the nation. A great deal of progress has been achieved in improving water quality through the regulatory mechanisms established by the CWA and through use of pollution prevention and other components of the second epoch of the 1980s and 1990s. Yet the area continues to struggle, as does the rest of the country, with controlling nonpoint sources of water pollution (where the origins are multiple, varied, and diffuse) and with removing highly dangerous toxic chemicals from its streams, rivers, lakes, and bays. Doing so is essential to promote community sustainability in a region where the culture and economic vitality are tied firmly to maintenance of a high level of water quality. Recent efforts in the region to protect water quality have depended heavily on the use of collaborative decision making and public-private partnerships. The successes and limitations of these approaches hold some lessons for the rest of the nation.

Water Quality Problems and Progress

Water resources and their quality are vital to life and to the nation's economy. Water resources meet the public's need for clean drinking water, and they also support agriculture, industry, electric power generation, recreation and tourism, transportation, and fisheries. In the Fox-Wolf Basin, clean surface waters are the primary source of drinking water for about 100,000 people. Industries throughout the region use water for papermaking and food processing, among many other industrial activities. High-quality surface water also helps to satisfy the public's diverse recreational interests—fishing, swimming, canoeing, sailing, boating, and the simple enjoyment of the beauty of a large body of water flowing through a wooded countryside or the middle of an urban area. The river basin and bay of Green Bay are also vital to the integrity of the Great Lakes ecosystem; they are a key component.

Degradation of Water Quality

The quality of the nation's water is affected by a multiplicity of human uses, including point discharges of waste from industry and municipalities and nonpoint source pollution from agriculture and urban runoff. A century and more of industrial development, burgeoning human population, and land use changes throughout the nation had taken its toll by the time the Clean Water Act was adopted in 1972. The waters of the Fox-Wolf Basin were no exception.

The lower Fox River stretches 39 miles from its source near the cities of Neenah and Menasha in the northwest corner of Lake Winnebago north through the Fox Cities to its mouth at the head of the bay of Green Bay. The lower Fox River Basin encompasses approximately 400 square miles of drainage area in Northeastern Wisconsin. The downstream end of the Fox-Wolf River Basin drains about 6,400 square miles of land in Northeastern and East Central Wisconsin (see figure 4.1) and supports a population of about 750,000 people. The area has the highest population growth rate in the state outside of Madison. The Fox-Wolf River Basin is the largest tributary to the Lake Michigan Basin and the third largest to the Great Lakes. It also empties into the head of the world's largest freshwater estuary in the bay of Green Bay. The basin is a complex ecosystem that has been transformed substantially over the past century, especially since the 1920s. It is important too, because problems in the basin mirror those in the larger Great Lakes region.

The lands and waters of the Fox-Wolf Basin historically provided a wealth of resources for its inhabitants. Native American tribes foraged for fish, wildlife, and grains for centuries prior to European settlement. Following European exploration, the area's abundant natural resources, especially its water, fueled intensive economic development.

As a consequence of that development, water quality began to decline as early as the 1830s, when extensive clear cutting for lumber of 300-year-old white pine and maple began on lands formerly belonging to the Menominee Indian Tribe. As lumber industry activities decreased in the lower Fox River Valley, having exhausted the local supply of suitable trees, pulp and paper mills grew, making use of new harvests of poplar, birch, and other trees from more distant areas in the state. By the 1920s, thirty-four pulp and paper mills had been established along the lower Fox. In the 1990s, the lower Fox received the discharge of thirteen mills, the largest concentration of the paper industry in the world. The river and bay have suffered from their prodigious discharge of a variety of waste products. Massive fish kills that began in the 1920s signaled what lay ahead for the river.

In addition to the paper mills, area water quality was adversely affected by the operation of metal fabricators, tanneries, breweries, food process-ing mills, and a host of other large and small businesses that directly dumped their wastes into area streams and rivers. Municipal sewage plants had long discharged insufficiently treated human and other waste before their treatment systems were modernized in the 1970s and 1980s. Also contributing to the deterioration of water quality in the region was

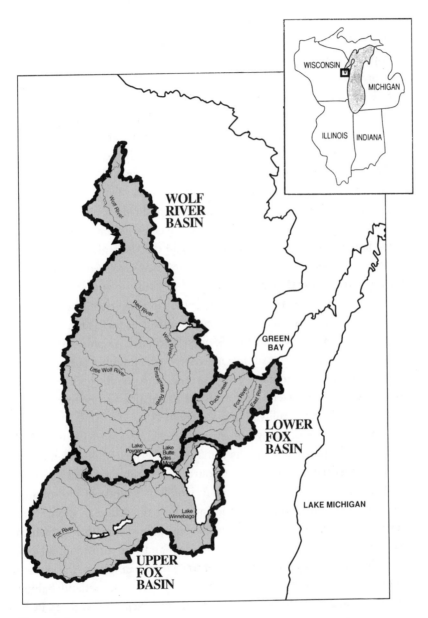

Figure 4.1
The Fox-Wolf River Basin

the loss of more than 90 percent of coastal marshes (wetlands) in the lower Fox River and Green Bay to land filling, industrial and residential development, and recent high lake levels, some of which are permanent and represent irreversible damage to the ecosystem (WDNR 1993; Bosley 1978; Harris, et al. 1982).

Beyond the paper mills and other industrial and municipal sources, another major contributor to water pollution in the area has been agriculture, including a dairy industry that is heavily represented in the Fox-Wolf Basin. Agricultural development transformed the landscape and significantly increased the runoff of soils, nutrients from fertilizers and untreated animal wastes, and chemicals from pesticide applications. Adding to the chemical soup entering the river basin was runoff from urban streets, lawns, and construction sites throughout the region.

One of the most serious water quality problems, related to the paper mills and other industries but also rooted in land use and air pollution, is persistent toxic chemicals. Varying levels of more than one hundred potentially toxic substances have been identified in the water, fish, and sediment of the Fox River, including polychlorinated organic compounds such as PCBs, dioxins, and furans; mercury, lead, and other heavy metals; pesticides; polycyclic aromatic hydrocarbons; and volatile hydrocarbons. Most have not been adequately assessed for health and environmental effects (WDNR 1993; Sullivan and Delfino 1982).

Elaborate studies of ecosystem stressors in the region have identified nutrient and sediment loading as a major concern, resulting in the loss of many beneficial uses of water resources (WDNR 1988; Harris et al. 1994). In recent years, an estimated 1.5 million pounds of phosphorus have flowed into the Fox River and into the bay each year, along with an estimated 150,000 tons of suspended solids, the vast majority of which have come from nonpoint sources—algae and soil erosion (WDNR 1993). Water quality experts agree that to restore area waters, the loading of nutrients and solids must be significantly reduced, wetland habitat protected and restored, and persistent organic chemicals such as PCBs eliminated or reduced to a level where no adverse effects on the ecosystem can be detected (WDNR 1988; Harris 1994).

Persistent organic chemicals, such as PCBs, and other toxic chemicals, such as mercury and ammonia, remain the substances of greatest concern in the lower Fox River and Green Bay. All pose a risk to ecosystem functions in area waters. Toxic chemicals that are persistent and bioaccumulate also present a risk to human health, and both types of chemicals are very difficult, or impossible, to manage once they enter the environment (Harris et al. 1994; WDNR 1993). Between 1954 and 1997, seven area paper mills discharged an estimated 420,000 to 825,000 pounds of PCBs into the river, of which about 63,000 pounds remain today in the river's sediments; the remainder have washed into the bay. Some quantities continue to be discharged by point sources, but well over 90 percent of the PCBs entering the bay now come from sediment resuspension and settling (WDNR 1993 and 1999). Although many area residents do not realize this, the federal government considers the lower Fox River and Green Bay to be among the most severely PCB-contaminated sites in the country.

Progress in Improving Water Quality

In its twenty-fifth anniversary issue, the federal Council on Environmental Quality (1997, 12) noted that across the nation "most of the conspicuous water pollution from point sources has been eliminated. More than 57,000 industrial facilities now operate under a pollution control permit." Other observers would offer a less sanguine view of water quality, in part because of the paucity of reliable data about the thousands of streams, rivers, ponds, and lakes in the nation (Knopman and Smith 1993; Davies and Mazurek 1998).

Some trends are reasonably clear, however. Almost everywhere there has been a major reduction in the raw pollution of surface waters. The proportion of the U.S. population served by wastewater treatment systems rose from 42 percent in 1970 to about 74 percent by 1985, with a resulting, estimated decline in annual releases of organic wastes of about 46 percent (Adler 1993). Most of the huge financial investment in clean water since 1972 has been expended on conventional point sources of water pollutants, and as a result, most industries and municipalities have greatly reduced their discharges, consistent with the intent of the Clean Water Act.

Beyond those controls on dischargers, we have other ways to measure progress. The most widely used assessment of surface water quality is the

National Water Quality Inventory, a biennial report from the EPA that is based on inventory by the fifty states of their own water resources. Section 305(b) of the CWA requires states to report every two years to the EPA on the extent to which state waters support designated beneficial uses. The review of 1996 was based on surveys within the states in 1994 and 1995, with varying methods of assessment. Caution is usually urged in interpreting the numbers, however.[1] For example, the states surveyed only 19 percent of the nation's total miles of all streams and rivers (although 53 percent of those that flow year around), and 40 percent of its lakes.

Based on these limited inventories, for the nation as a whole, about 64 percent of rivers and streams and 61 percent of lakes, reservoirs, and ponds "fully support" such "designated uses" as swimming, fishing, drinking water supply, and support of aquatic life. About 36 percent of rivers and streams were found to be impaired to some degree as were 39 percent of lakes, ponds, and reservoirs. Nearly all Great Lakes shoreline areas were assessed by the states, and they found only 3 percent of them to fully support designated uses. About 97 percent were rated by the states as fair or poor, largely because of fish consumption advisories and unfavorable conditions for supporting aquatic life—chiefly because of persistent toxic chemicals that enter the food web, habitat degradation, and competition by nonnative species (U.S. EPA 1998).

These numbers show some improvement over previous years, yet they also indicate that many problems remain. Their causes are also fairly evident. EPA reports during the 1990s have indicated that the leading sources of water quality impairment related to human activities affecting rivers have been (in order of importance): agriculture, municipal sewage treatment plants, hydrologic and habitat modification, urban run-off/storm sewers, and resource extraction. That is, the remaining problems are largely nonpoint sources of pollution. Neither Congress nor the EPA has done much to address such sources, although the state of Wisconsin has made some progress, as noted later. The EPA has had a modest program to aid the states in addressing nonpoint sources ($95 million per year in grant funds), but as critics note, this amount is less than 5 percent of what the agency spends on water quality programs (Davies and Mazurek 1998, 272).

Cleaning up the Fox River

Improvements in water quality are as apparent in the Fox River and Northeastern Wisconsin as they are nationwide, and these achievements reflect the stringent regulatory framework created by the Clean Water Act in a manner that was typical of environmental protection policy in epoch one. Until passage of the act in 1972, the lower Fox River Basin remained heavily polluted despite many state and local attempts to clean it up. After passage, efforts were far more successful. By the middle to late 1970s, fish and other organisms began to return to the river as wastewater treatment facilities were built or modernized—in part through funds made available by the CWA. By the early 1980s, a world-class walleye fishery had been restored in the seven mile stretch from De Pere to the mouth of the Fox River in the bay of Green Bay.

Those gains in water quality were achieved mainly through the state-administered permitting system that sought to meet water quality standards set by the CWA. The permits limited overall pollutant discharge in the entire 39-mile stretch of the lower Fox by allowing specified discharges only to the extent that the river was capable of assimilating them. Under Section 402 of the CWA, states implement the National Pollutant Discharge Elimination System (NPDES), contingent upon approval by the EPA. Under Section 307 of the act, the NPDES also has been used to control discharge of PCBs and other toxic chemicals.

The results of this regulatory action were striking. According to one recent assessment, from 1971 to 1990, total suspended solids discharged by point sources to the Fox River declined by 91 percent. Between 1962 and 1990, biological oxygen demand loadings (a key determinant of river health) fell by 94 percent. The author concluded that "up and down the river, the story is much the same. The cleanup of industrial and municipal discharges over the past three decades has been nothing short of astonishing" (Alesch 1997, 8).

A local journalist captured the same sentiment, with a sober warning of work yet to be done:

Through the diligence of environmentalists, the force of regulators and the cooperation—sometimes reluctant—of paper mills, great strides have been made in cleaning the water in the past twenty-five years. In the summer, fishermen line its banks and boats ply its shimmering waters. Many fish and fowl, once driven out by poison and pollution, are coming home. But it's still a sick river. The Fox remains unfit for swimming and offers up more than a dozen fish species that are unsafe for human consumption. Visitors come to look but not touch. (Campbell 1997d)

As is the case nationally, the remaining problems in the Fox-Wolf River Basin are largely nonpoint sources of pollution, especially nutrients and sediments, and toxic chemicals. Many of those chemicals, especially PCBs, are buried in the river's sediments, and "leak" into the river through resuspension and biological uptake. They will be a major object of concern for decades to come.

A Shift in Emphasis: Cost-Effectiveness and Collaboration as Guiding Principles

As this brief assessment of progress under the Clean Water Act illustrates, the regulatory apparatus established under the act, representative of the first epoch in environmental policy, could go only so far. The laws and regulations could yield impressive gains in many areas with a high level of political commitment and strong state capacity for administration of the programs (Ringquist 1993). Yet the CWA also faced major barriers because of the high cost associated with meeting stringent water quality standards, opposition from industry and municipalities, and continued contribution to water quality problems of largely uncontrolled nonpoint sources of pollutants. Hence the appeal of the efficiency-based regulatory reforms and collaborative approaches to decision making associated with epoch two. The latter is especially promising as a new policy strategy even if not fully tested.

Both nationally and locally, proponents of collaborative decision making argue that it can replace the often highly adversarial relations between government and industry and between the federal and state governments (National Academy of Public Administration 1995; Weber 1998a and 1998b). Collaboration involves greater participation by a diversity of stakeholders, building of trust, and a freer exchange of information, which together are expected to lead to a more open decision-making process and a greater commitment to achieving environmental quality goals. Its advocates believe it will yield more successful policies and programs, measured by achievement of improved environmental conditions at a faster pace and lower cost than what is likely under conventional regulation.[2]

As is true of most new approaches to environmental policy, however, such expectations are sometimes grounded more in dissatisfaction with past policies and faith in the new efforts than they are in a successful track record. Historically there have been few systematic evaluations of the EPA's environmental programs, and there is even less evidence of how well the new approaches of epoch two have been working (Knaap and Kim 1998; U.S. General Accounting Office 1997a and 1997b). What remains uncertain is the ease with which such a collaborative style of decision making can be established and maintained, and the extent to which it will actually help to achieve both the desired environmental quality goals and community sustainability.

Collaboration, for example, may work best in the early stages of the policy process—in the identification of problems, the consideration of alternative policy approaches, and the selection of the tools to be used. Such cooperative approaches may be less suitable, however, for implementation of the chosen policies—when specific action steps must be taken. Implementation success may depend on the existence of sufficient authority at the state or federal level to maintain timely progress in reaching environmental goals. Assessment of experience to date in the Fox-Wolf Basin helps to illuminate these relationships.

Emphasizing Cost-Effectiveness in the Fox-Wolf Basin
The new emphasis on cost-effectiveness as a policy strategy can be seen in the efforts of local organizations in the Fox-Wolf River Basin during the 1990s. Industries and municipalities along the lower Fox River had invested a great deal of money in technological pollution controls under the Clean Water Act, spending an estimated $300 million by 1987. The Green Bay Metropolitan Sewerage District (GBMSD) spent a considerable portion of that amount, and it was as eager as industry to find ways to minimize future costs. Both industries and municipal dischargers pressed increasingly for a different approach, and cost-effectiveness became their watchword. They were convinced that additional investments simply could not be justified on the basis of the expected benefits in water

quality, which they believed could be improved only marginally if at all (Alesch 1997). They worried about how to comply with the additional controls they expected the state to impose within a decade as it continued on course to implement the CWA. They also were concerned about the likely impact on an area economy that is heavily dependent on paper making, paper recycling, meat packing, and vegetable processing (all water intensive).

In 1992 government and industry leaders established a not-for-profit corporation, Northeastern Wisconsin Waters of Tomorrow (NEWWT) to assist in developing a new strategy. NEWWT was funded in large part by the municipal sewerage district, which had assumed a major role in actions on local water pollution. The GBMSD was determined to find alternatives to expensive treatment of water problems by the district itself, the cost of which it would have to pass on to area residents. Its new relationship with NEWWT provided a possible avenue for developing such solutions. NEWWT's purpose was to search for cost-effective alternatives to conventional regulation in the watershed. It did so with the blessings of the Wisconsin Department of Natural Resources and its director, George Meyer, an appointee of the longtime Republican governor, Tommy Thompson.

NEWWT recruited a small staff, hired consultants, and developed a computer simulation of the Fox-Wolf River Basin. Of particular concern for dealing with the problem of phosphorus and suspended solids in the river basin was how interdisciplinary analysis of the watershed could improve understanding of the flow of water, nutrients, solids, and other material that affect the quality of the river and bay. Such knowledge also could help in setting pollution control targets to maximize water quality improvements. In short, where should citizens, industry, and the state invest money and energy to have the greatest impact on cleaning up the river and bay?

Using existing data, the research team concluded that about 75 percent of the phosphorus and 90 percent of the suspended solids that reach lower Green Bay came from rural upstream sources. Economic analysis also indicated that the cost of controlling those sources would be significantly less than the alternatives: $9.64 per pound for agriculture compared with $165 per pound to reduce phosphorus in municipal treatment facilities; and $.008 (less than a penny) per pound for preventing suspended solids at the outlet of a watershed compared with $4.61 per pound at the treatment plant (White, Baumgart, and Johnson 1995).

The message could not be clearer. The biggest payoff for the least cost would come through reducing phosphorus and solids from agricultural lands entering the waters. These objectives could be met through planting of vegetation along river banks ("vegetative buffers") to minimize erosion, encouraging more efficient use of fertilizers and pesticides, adoption of zoning and livestock exclusion ordinances to minimize animal access to area streams, and improved management of animal wastes, among other actions. The state DNR has actively pursued these goals through educational outreach programs directed at changing farming practices (Wisconsin agriculture consists largely of small family farms), provision of technical assistance, and partial subsidy of the costs of new nonpoint source controls.

Such efforts to address nonpoint sources of water pollution underscore the long-term and difficult challenge of improving water quality in the basin. Applying technology to "fix" the point discharges, the key focus of epoch one in environmental policy, can never be sufficient, and it cannot be economically efficient to continue to ratchet up the regulatory requirements on those sources. Yet dealing with the nonpoint sources will take longer and involves working with a greater diversity of people, economic enterprises, and landscapes. A recent report on the state of the bay tried to respond to the understandable public frustration with the painfully slow process of restoring area water quality based on this new watershed approach:

[T]here are no "fix-it" manuals for large scale ecosystems, and the economic and social matrix of our society resists change. It is not a matter of applying the right technological fix, but rather a matter of changing the whole way we count environmental costs and measure the effectiveness of expenditures. We are in that process and the learning curve is steep. It will take time to mobilize resources appropriate to the scale that is needed. The results, however, should be sustainable because we will come to see ourselves as part of a larger system, the "health" of which is necessary for our own health and economic well-being. (Harris 1994, 24)

NEWWT and GBMSD Favor Collaboration as a Strategy

The search for cost-effective solutions to local water quality problems (a key feature of epoch two in environmental policy) pushed NEWWT as well as the GBMSD to adopt a waterbasin or ecosystem perspective. In doing so, both built on previous efforts by other local groups and the WDNR itself to use an ecosystem approach in dealing with water quality problems in the area (H. Harris, et al. 1982; V. Harris 1992). To better reflect this orientation, NEWWT changed its name to Fox-Wolf Basin 2000 (FWB 2000), added members to its board of directors from across the watershed, and became the most active proponent in the region of an ecosystem approach to improving water quality. Although in some respects FWB 2000 is a grassroots environmental organization, it is also self-consciously dedicated to a collaborative philosophy of working closely with business and government agencies, which distinguishes it from many other area environmental advocacy groups. Some of the latter have been vocal opponents of FWB 2000, which they consider to be too closely aligned with area polluters.

At about the same time it helped to initiate FWB 2000, the GBMSD entered into partnerships with area businesses to promote its new philosophy of pollution prevention and cost-effective approaches to water quality improvement. For example, the GBMSD cooperated with the county's solid waste department to create the state's first full-time center for collection of household hazardous waste that might otherwise have found its way into area waters. The GBMSD generally found a supportive business community that was seeking new ways to enhance its competitiveness in a global economy. Hence it was attracted to cost-effective ways to control area pollution, or to prevent it through changes in production processes (Alesch 1997).

In addition to its many other actions, the GBMSD (with help from Fox-Wolf Basin 2000) prodded the Wisconsin DNR to move more aggressively on its long-standing commitment to using a watershed approach to environmental protection, and to use the case of Northeastern Wisconsin as a pilot program for reorganization of the WDNR itself along ecosystem lines. The new approach parallels efforts by the Clinton/Gore administration to "reinvent" environmental regulation. It is too early to anticipate how successful the effort will be, but the reorganization hints at what could become one of the most fundamental transformations of environmental management within the state in decades.

These changes represent the beginning of a transition from what we have called epoch two to the sustainability-based focus of epoch three.

Indeed, DNR Secretary George Meyer himself has referred to the organizational shift as a "new paradigm," and he has linked it directly to other efforts within the state to "cut red tape and costs" and to find alternatives to "the cumbersome, burdensome, and expensive regulatory process" (quoted in Alesch 1997, 18). Meyer suggested that the department, which is the designated agency for both pollution regulation and management of natural resources in Wisconsin, had few choices in a period in which flat budgets left it unable to get its work done without entering into such partnerships. These partnerships were attractive in part because the new partners were expected to add to the total funds available for water quality initiatives. Yet it also became evident that WDNR's professional staff (especially those working in Madison, the state capital) were considerably less enamored of the organizational changes being promoted, and they resisted it. Many environmentalists worry as well that the regulatory process itself, and public and environmental health, might be compromised now that the WDNR secretary is appointed by the governor; previously an independent Natural Resources Board selected the secretary.

One of the most intriguing questions raised about these recent developments is whether a regulatory agency can successfully negotiate such organizational and policy changes without sacrificing the very qualities that have led to improvements in water quality since the early 1970s. That is one of the risks of de-emphasizing regulatory approaches in search of more cooperative (and more politically appealing) relations in environmental policy. There is much less question that adoption of a watershed approach and a strategy of ecosystem management represents a major advancement over previous water quality regulation within the state. The federal EPA has been encouraging such an approach across the nation, and it is representative of epoch three thinking.

The Remedial Action Plan Process as Collaborative Decision Making

One of the earliest efforts at collaborative decision making was well under way by the time the WDNR was considering reinventing itself. This venture grew out of the Great Lakes Water Quality Agreement signed by the United States and Canada in 1972, and amended in 1978 and 1987 (Botts and Muldoon 1997; Colborn et al. 1990). The agreement highlighted the need to clean up persistently polluted trouble spots, or Areas of Concern (AOCs), throughout the Great Lakes Basin, largely in ports, harbors, and river mouths that are tributaries to the Great Lakes. The International Joint Commission (IJC), the American and Canadian organization that oversees binational interests in the Great Lakes, identified forty-three such AOCs in the basin. One of these was the lower Green Bay and the lower Fox River downstream of the De Pere dam. The agreement also recommends that states and provinces prepare Remedial Action Plans (RAPs) to define corrective measures and long-term strategies to remedy persistent environmental problems that had impaired use of the tributaries that feed into the Great Lakes. The development of such plans takes place through a process of collaboration among a diversity of stakeholders.

The Green Bay RAP: Organization and Purposes

The Green Bay RAP, like the others throughout the basin, is a set of recommendations and guidelines, a "community plan," developed by the WDNR in "partnership with local governments, other agencies, businesses, and many interest groups—agricultural, academic, environmental, conservation, and recreational—in the Fox-Wolf River Basin" (WDNR 1993, xiv). Work on the RAP began in 1985, when the WDNR initiated the development of a plan for Green Bay, which, as noted, had been designated as an Area of Concern (AOC). A Citizens Advisory Committee and four technical advisory committees contributed to the plan, which was completed in 1987, and approved by the state as part of its Water Quality Management Plan in February 1988. Because the Lower Green Bay RAP was one of the first within the Great Lakes Basin to be completed and the first to receive approval from the IJC and its Water Quality Board, it has been cited widely in the region as a model for RAP development and implementation. It also has been praised for using citizen involvement and an ecosystem approach to planning and management for water quality improvement (Harris 1992).[3]

The RAP is, among other things, a vision of the future. That vision is based on a broad ecosystem view, expressed in both general goals and objectives and specific recommendations to restore ecosystem functions. Achieving those goals would help to realize a "desired future state," in which the full beneficial uses of area waters long impaired by human activities would be restored. As such, the RAP process in many ways epitomizes the policy approaches characteristic of epoch two—and emerging in the sustainable community model of epoch three.

Meeting these various goals and objectives required that the RAP develop and use extensive modeling of the river and bay and ecological risk assessments to determine how specific stressors affect water quality and aquatic life. Success of the plan also required development of appropriate indicators of local ecosystem health that could facilitate long-term monitoring to determine whether the system was responding to the remedial actions undertaken. A "State of the Bay" scorecard was developed for this purpose, with special attention given to indicators that were consistent with scientific investigations of area waters, yet could be understood by the "informed public" and decision makers, and thus become part of the essential local dialogue over remedial activity (Harris 1994; Harris and Scheberle 1998).

RAP Recommendations and Achievements

These institutional processes produced a set of long-term water quality goals that were reported in the original RAP of 1988, and in an updated plan released in 1993. The endorsed goals reflected concern not only for ecosystem health but also for social and economic conditions in the area. Thus they incorporated at least some of the expectations for building sustainable communities. These goals included: a healthy bay environment; a balanced, edible sport and commercial fishery; productive wildlife and plant communities; water-based recreational opportunities; good water quality that protects human health and wildlife; balanced shoreline uses; and an environmentally sound and economical transportation network.[4] The 1993 update of the plan added an eighth goal that made the link to sustainability much more explicit: "to ensure the sustainability of a restored and healthy environment through pollution prevention and the development of sustainable economies, resources, and facilities which support beneficial uses into the future" (WDNR 1993, xxii).

The technical committees of the RAP developed 120 detailed recommendations for achieving these general goals and objectives, which were reviewed at two public information meetings and a public hearing. The

Box 4.1
The highest-priority actions recommended by the Lower Green Bay Remedial
Action Plan

• Eliminate toxicity of industrial, municipal, and other point source discharges.
• Reduce availability of toxic chemicals from contaminated sediments.
• Reduce phosphorus inputs to the river and bay from both point and nonpoint sources.
• Reduce inputs of sediment and suspended solids.
• Create an institutional structure to implement the plan.
• Increase public awareness of, participation in, and support for the river and bay cleanup (WDNR 1993).

---

highest priority actions that emerged from this process are listed in box
4.1.

Between 1988 and 1993, some 38 of the 120 recommended remedial
actions were implemented. Another 57 were initiated, and 25 had not yet
begun. As might be expected, most of the actions completed were relatively short-term and inexpensive projects such as voluntary reductions
in phosphorus discharges, public access improvements, habitat rehabilitation, native fish reintroduction, and nonpoint source management demonstration projects. More significantly, the RAP drew attention and
resources to the problem of runoff pollution, which resulted in about a
dozen large scale nonpoint source management projects in the basin over
the past decade, including three priority watershed projects. The WDNR
made available some $50 million to $100 million in cost-sharing funds
to help communities and landowners meet those goals of reducing nonpoint sources of pollution.

As these examples illustrate, the RAP's strength has been in bringing
diverse interest groups together to define the problems and solutions, to
help set priorities among competing problems, and to serve as a catalyst
for action. It also facilitated the integration of scientific findings and
actions by both public- and private-sector parties, fostered and supported
public education on the issues, and facilitated public participation in
environmental decision making (Harris 1992).

Despite these important achievements, however, it was widely understood that existing statutes and resources would be insufficient to achieve the water quality goals set out in the RAP recommendations. As the 1993 RAP update made clear, "loads of toxic contaminants from point sources appear to be largely controlled, but nonpoint sources, including contaminated sediments [were] not reduced significantly from 1985 and are substantial, continuing pollutant sources to the AOC" (WDNR 1993, xxi). The RAP update acknowledged that restoration of area waters will take decades of sustained commitment to pollution prevention, cleanup of contaminants, habitat enhancement, and better land use management, among other actions. Many critics were skeptical that the commitment and restoration actions would be forthcoming, in part because the RAP as an institutional entity has no formal authority to bring about such change. Rather its recommendations are intended to stimulate citizen concern and public- and private-sector responses to water quality problems in the region (Harris 1992).

The Problem of Contaminated Sediments: Can Collaboration Work?

Among the most challenging of the tasks identified by the RAP is the cleanup of contaminated sediments in the lower Fox River, which is expected to reduce the flow of persistent toxic chemicals into the bay and into Lake Michigan. The chief concern is large deposits of PCBs in the river sediments. The sources of the problem, the key health and environmental effects, and implications for cleanup are summarized in box 4.2.

Contaminated sediments are common to the 43 AOCs in the Great Lakes Basin, and significant progress has been made in several locations within the basin in dealing with them. In Green Bay, however, debate over the extent of cleanup necessary and who was to pay the exceptionally high cost had become so fierce that conventional regulatory processes alone had little chance of success. As was the case with achieving cost-effective cleanup of the river through targeting of nonpoint sources of pollution, removal of contaminated sediments seemed a good candidate for the use of collaborative decision making.

According to the Green Bay Mass Balance Study and other recent investigations, more than fifty contaminated sediment deposits exist along the lower Fox River.[5] Most of these are just downstream of industrial outfalls. Each deposit is unique and its effect depends on hydrologic conditions, the number and types of contaminants, the extent of

Box 4.2
The problem of PCB contaminated sediments

First manufactured in 1929, PCBs were used in numerous industrial processes because of their relatively low cost and stability. They found their way into electrical equipment, hydraulic fluids, and other commercial and industrial products before their manufacture was banned by the federal government in 1976. PCBs became a problem for the Fox River largely because of their use in carbonless copy paper that was recycled by area de-inking paper mills (and produced by one of them) that lined the river. As a result, these mills routinely discharged PCBs to the river system before the practice was made illegal. The quantities were small compared with the total volume of discharges, but the PCBs accumulated over time.

By 1972, the use of PCBs in carbonless paper was banned, and regulations have since reduced direct PCB discharges to negligible levels. However, the PCBs, along with other toxic substances, remain attached to sediments in the river, and from there move into the water column and bay. Each year an estimated 600 pounds of PCBs from sediments are flushed from the lower Fox River, with a much larger amount removed in years of heavy rain and high river flows.

Numerous studies have confirmed the adverse impact of PCBs on aquatic life, and all waters in Green Bay and Lake Michigan exceed PCB water quality standards established to protect aquatic organisms and other wildlife. Human exposure to high levels of PCBs has been linked to increased risk of reproductive, neurobehavioral, and developmental effects (such as impaired responsiveness and reduced mental abilities in infants and children); liver damage; immune system effects; and cancer, among other health problems. Human exposure to PCBs and other potentially toxic substances in the Great Lakes Basin is predominately through eating of contaminated fish. The concentration of PCBs in area fish has led since the late 1970s to advisories cautioning human consumption of various species, particularly for pregnant mothers and infants. Consumption of similar amounts of mercury contaminated fish can lead to central nervous system suppression and neurotoxicity. Such health risks remain uncertain, however, contributing to debates over the acceptable level of risk related to continuing water quality problems. In 1998, the WDNR conducted a new risk assessment funded with $1.65 million from the U.S. EPA. Using extensive data collected previously, the study indicated that PCBs in the lower Fox River pose a health risk from 100 times to 1,000 times greater than from any other chemical contaminant present in the water, including dioxins, furans, DDT, dieldrin, arsenic, lead, and mercury.

The adverse impacts of the PCBs are likely to diminish over time, but only very slowly. Recent estimates from an exhaustive EPA and WDNR joint study of PCBs in the river and bay indicate that without remediation

Box 4.2
(cont.)

of the contaminated sediments, the impairment of water quality and the associated ecological effects may continue for 120 to 150 years, or more. Even minute amounts of PCBs in the water can adversely affect wildlife (and humans) because they concentrate through biomagnification as they move up the food chain—by 10,000 times or more—and they are highly persistent. Impaired uses could be restored, but it is not physically possible to remove all of the PCBs in the system, and substantial quantities of sediments cannot be removed quickly. Even with massive remediation efforts (at a very high cost), recovery is likely to take many decades.

Sources: WDNR 1993 and 1999; Smith et al. 1988; Beyer, Heinz, and Redmon-Norwood 1996.

contamination, the constitution of the sediment, and the deposit's depth, breadth, and location. The high degree of variability means that different forms of treatment or remediation may be necessary at different sites, with sharply varying costs (Harris 1994; WDNR 1993). Most of the PCBs will remain buried in the river, but depending on aquatic conditions, some portion enters the water column or is volatilized into the air and transported downstream or downwind. Because some of the deposits are at significant risk of movement, the timing of their remediation is an important issue. Once the contaminants reach the bay of Green Bay or Lake Michigan, there is no feasible way to remove or remediate them.

The conflict over what to do about contaminated sediments lies at the heart of water quality policy debates in the region in the late 1990s. It is within this context that government officials and business groups grew highly skeptical of the conventional regulatory framework and began promoting the alternative of collaborative decision making. The modest success of this model of decision making in the development of the RAP may have contributed to its appeal. The operations of the Public Advisory Committee of the RAP (the leading entity in that process) and the RAP's several other committees demonstrated that business leaders, government officials, environmentalists, community leaders, and academics could work together and build consensus on water quality problems and develop an action agenda. Similar cooperation between businesses and municipalities from the mid-1970s to the mid-1980s in developing a waste-load allocation model (for biological oxygen demand) for the lower Fox River also suggested the feasibility locally of such a model of cooperation and consensus building for the contaminated sediment problem. There were significant cost savings to both industries and municipalities as a result of this effort, which took place under provisions of Section 208 of the Clean Water Act.

The constraints on conventional regulation in dealing with contaminated sediments are evident to all concerned in the Fox-Wolf Basin. Elaborate and comprehensive scientific investigations cannot eliminate the inherent uncertainty of risks to human and ecological health. State and local governments, as well as industry, have limited financial resources available for cleanup of contaminated sediments, and they are reluctant to act without clearer evidence of the benefits that could be expected. As is often the case with remediation of hazardous waste sites, sharp disagreement exists over what constitutes a fair allocation of the cleanup costs. There are no easily available and proven remediation technologies that can be used on contaminated sediments, and the most technically attractive (removal and land disposal or incineration) are both costly and likely to face stiff public opposition. The intricate labyrinth of federal and state regulatory policies for removal and destruction of contaminants creates an almost insurmountable task for those charged with improving and protecting the region's water resources. Under such conditions, the appeal of collaboration is understandable, and such a strategy was pursued through a newly formed Fox River Coalition.

Collaboration and River Cleanup: The Fox River Coalition

The idea for a Fox River Coalition (FRC) began emerging in late 1991, as a response to the increasingly intractable problem of contaminated sediments and the legal morass local businesses faced under the federal Comprehensive Environmental Response, Compensation, and Liability Act (CERCLA), better known as Superfund. The law requires action to clean up such contaminants, and it authorizes the U.S. EPA to compel responsible parties to take action or to pay for remediation. The paper mills developed an intense interest in seeking voluntary cooperation to deal with contaminants in the hope that they could steer remediation toward cost-effective strategies and a combination of public and private funding. This approach was attractive as one way to avoid what could otherwise be a very costly damage assessment. The mills also pressed continually for further scientific studies to help ensure, in their view, that solutions would be technically justifiable and cleanups not more stringent than necessary. The FRC was formally established in July 1992, to further these goals (Mercurio 1995). Local government officials shared many of the concerns voiced by the pulp and paper industry over cleanup costs and the adequacy of available scientific data for sound decision making.

The coalition's membership included about a dozen paper mills located along the lower Fox, the WDNR, and local governments with publicly owned water treatment works. Altogether, some 30 "partners" crossed political boundaries to explore the technical, financial, and institutional needs for comprehensive remediation of the sediments. Central to these discussions, including those at the highest level in the state—involving the WDNR secretary and members of the Wisconsin Paper Council's Executive Committee—was the idea that a cooperative approach for remediation of the Fox River's contaminated sediments would be beneficial to all concerned (Johnson 1996).

Much of the work of the FRC focused on review of available data and information gaps, and education of coalition members (many of whom were managers or public officials with no scientific background) on the technical aspects of remediation. A Technical Work Group of the FRC drew members as well from the RAP's Science and Technical Advisory Committee's work group on contaminated sediments. The FRC work group contracted for detailed assessments of selected deposits in the river, and evaluation of options for cleanup of the sites. Industry paid for about 25 percent of the cost, with state and local governments covering the rest, a sore point with some members, particularly in anticipation of how the costs of the eventual cleanup itself would be allocated. Cost estimates for cleanup of the entire river system ran from zero (if no action were taken) to over $2 billion, based on comparable sites elsewhere in the Great Lakes.

Throughout this first phase of the cooperative approach, little attention was paid to establishing formal procedures regarding FRC membership and the conduct of meetings. The process bore a strong resemblance to the "garbage can" model of organizational decision making described by Cohen, March, and Olsen (1972), with inconsistent and ill-defined preferences; unclear technology; decisions made by trial and error; and fluid participation, or substantial variation in time and effort devoted to the tasks by participants.

Consistent with this characterization, the FRC made what may have been a serious strategic error. It decided early in its history against inviting environmental and conservation organizations to participate, even though they had been well represented in the earlier RAP process. Minutes of the coalition's meetings indicate that it sought to limit participation to achieve "fruitful discussion and decision making" (Johnson 1996).

Discussion of issues at FRC meetings was usually informal, and decision making was largely by consensus, or lack of contention. There were few recorded votes at any of the FRC meetings, which were usually chaired by upper-level WDNR staff and sometimes by the director of the WDNR's Water Resources Bureau. The WDNR itself handled the agenda, meeting summaries, and other administrative matters. The style of decision making followed by the FRC, and by the RAP process as well, which was procedurally similar, has many advantages, particularly in building consensus on agendas and actions plans. Yet there is a significant risk as well. Almost any member can "veto" a proposal through a strongly voiced objection, as long as others have no equally firm conviction on the issue. Such an approach may keep potentially attractive policy alternatives off the table. It may also contribute to what critics see as a seemingly endless discussion of issues that produces no definitive decisions to take action.

The WDNR's summary of the coalition's actions describe it as a "national model for successful environmental restoration" and "Wisconsin's greatest hope for achieving successful river restoration in the shortest time possible" (Mercurio 1995, ii). Even the coalition's most fervent supporters, however, concede that progress was slow at the same time that "the mold was cast for a cooperative, nonadversarial approach to the problem" (Alesch 1997, 20).

By early 1994, this approach seemed to be going well, albeit slowly, until the U.S. Fish and Wildlife Service proposed a dramatically different policy strategy for dealing with the contaminated sediment problem. The FWS decided to pursue a Natural Resource Damage Assessment (NRDA), an investigation, evaluation, and quantification of injuries to natural resources that is conducted with the goal of calculating monetary damages to be used in restoration of the resources. Such action is authorized under the federal Superfund law for "trustees" of the nation's natural resources. The agency's decision to rely on this NRDA process posed a direct challenge to the ongoing work of the FRC.[6]

Limits to Collaboration Appear: The NRDA Process

The threat of a damage assessment and litigation over cleanup actions nearly ended the FRC process of collaboration and consensus building. Curiously, the NRDA seemed to have caught FRC members by surprise, despite common knowledge that it was being considered. The NRDA was favored by local environmental groups as a more definitive way to resolve the issues and force cleanup of the river, although it was by no means clear that the NRDA could achieve those goals. Despite these uncertainties, the threat of a FWS lawsuit based on the damage assessment would compel the FRC to expand participation and become far more visible than it had been.

The NRDA process also prodded the FRC to move more deliberately on the technical tasks of assessing the sediment removal problem and available options (including a "no-action" option) and cost estimates for remediation. Yet the technical working group could make little headway without a clear understanding of just how much cleanup was to be undertaken. None of the FRC participants was prepared to define the desired "end points" without prior knowledge of the costs of achieving such objectives and an agreement about local sharing of costs. That condition contributed to a lack of focus and direction on the part of the technical staff. With no clarity on the ultimate objective of water quality, it would not be possible for the public and policy officials to make much sense of the vast amount of technical information being generated. By 1995, all of these actions came under increasing scrutiny as the work of the FRC became more public.

Some environmental groups grew indignant that decisions were being made in secret, and the FRC increasingly was viewed by its critics as making no progress at all in cleaning up contaminated sediments. Those groups looked very favorably upon the NRDA, and in April 1995, a broad coalition of them, including the Sierra Club, Northeast Wisconsin

Audubon Council, Citizens for a Better Environment, Wisconsin's Environmental Decade, Clean Water Action Council, and the Lake Michigan Federation, appealed to Department of Interior Secretary Bruce Babbitt (to whom the FWS reports) to continue the NRDA process as an alternative to the protracted discussions between industry and the state through the FRC.

Above all, the environmental groups sought a definitive plan to begin early cleanup of the sediments and to have the responsible parties pay most of the costs. They were quite vocal in denouncing a "taxpayer bailout" and "corporate welfare" that they associated with the broader sharing of costs favored by the FRC. The most vocal among them, the Clean Water Action Council, was strongly opposed to any plan that smacked of compromise on cleanup standards or deadlines. It also tended to characterize both industry and government players as unworthy of the public's trust. Most of the environmental groups, however, agreed with paper industry and regulators' concerns that legal wrangling be kept to a minimum and that cleanup of the river be as cost-effective as possible. Nonetheless, they expressed skepticism over collaboration between industry and government if the result was less vigorous enforcement of current environmental regulations.

Fox-Wolf Basin 2000, originally formed to promote and facilitate cost-effective approaches to land and water resource management in the region, was more directly involved with the FRC than other environmental groups. It was also more supportive of the concept of a voluntary, cooperative approach to contaminated sediment remediation. It took the position that the alternatives could only create further delays, direct scarce resources away from the cleanup effort, and contribute to additional contamination of the bay of Green Bay (Johnson 1996). That stance cost FWB 2000 the support of other environmental groups, who denounced the organization as a front for industry. However, FWB 2000 also favored continuation of the NRDA process, largely because of its promise of producing useful information that would be necessary to achieve high standards for restoration of the Fox River.

The Elusive Pursuit of Consensus

The pace of the NRDA process did not initially offer much promise of being any quicker than the typical Superfund cleanup action or the record of the FRC's progress. If the NRDA were challenged by the responsible parties, the process could well drag on for years. Backers of the NRDA hoped that as the assessment moved along to concrete findings of economic costs and liability, the responsible parties would enter into negotiations with the Fish and Wildlife Service. The FRC feared that the NRDA actions would undercut its own technical investigations and pursuit of cost-effective solutions to the contamination problem. Above all, its members were concerned that a comprehensive cleanup for which industry would be expected to pay the lion's share would create a devastating economic burden that could seriously weaken the region's otherwise robust economy.

A series of public hearings or "open houses" sponsored by the FRC in 1996 did little to assuage the concerns of environmental groups. They persisted in lobbying for the NRDA process and in denouncing the FRC. Some of the meetings were highly confrontational in tone, with posturing on all sides. As is often the case, the discourse at such events may have confused the public more than it enlightened them. The FRC also attempted to build public support through a variety of communication efforts and pledges to operate more in the open. These actions appeared to be largely ineffective.

Similarly, efforts by the FRC in 1995 to open its meetings to environmental activists and others met with mixed reactions. The Lake Michigan Federation, Sierra Club, and FWB 2000 assisted in some form with the public meetings. Other environmental groups, particularly the "deep greens," refused to join the FRC in any manner, and they called the entire process corrupt, insincere, and "an avoidance of enforcement of the law." They made the same points frequently at meetings they did attend throughout 1995, as "non-FRC participants" (Johnson 1996, 49).

As this summary indicates, what began as a collaborative effort between industry and government that initially showed great promise, eventually was unable to operate effectively in public. The FRC also continued to suffer from a lack of clear direction about desired end states for the river (which frustrated the technical modeling staff that was expected to study cleanup options). In addition, behind the scenes, the paper industry, municipalities, state representatives in Congress, and the WDNR lobbied fiercely against the NRDA, which they saw as a threat to the established process of cooperative decision making and unwarranted federal intrusion into decisions they believed should be handled at the state level (Campbell 1997a).

In their formal meetings, as well as in other gatherings, FRC members continued to display a pronounced tendency to debate at length the accuracy and adequacy of data and models, acceptable levels of cleanup, cost estimates, and formal legal authority for final decisions. An extended time frame for discussion of technical, economic, regulatory, and political aspects of cleanup actions is doubtless necessary for all the key players to feel comfortable with the final decisions. It is a form of policy legitimation that will assist in implementation of the remediation strategy chosen. Yet such actions appeared to lend credence to the environmentalists' criticism that collaborative decision making was merely another device that facilitated industry and government procrastination on sediment cleanup.

On a more positive note, the experience of the FRC facilitated the development of sufficient trust to support cooperative working relations among the paper mills, the municipal sewage treatment plants, the Oneida Indian Tribe (whose lands lie within the river basin), and state and local government officials. These relationships continue in the late 1990s despite the end of the formal FRC process. Similarly, FWB 2000 continues to educate area residents and policy actors on the imperative of cooperation and the need to overcome political turf battles to achieve high-surface water quality in the Fox-Wolf River Basin. By early 1997, the FRC did manage to reach an agreement to develop cleanup plans for two sediment deposits on the lower Fox. The coalition's members pledged $10 million to fund remediation at these two demonstration sites and to preserve habitat in the area. Remediation plans were nearly complete by late 1997, although cleanup at the first (and smaller) site did not begin until late November 1998.

Toward Agreement on River Cleanup

Even with the considerable limitations of collaborative decision making represented by the Fox River Coalition, tentative agreement on a cleanup plan for the Fox River was reached by July 1997. Ironically, it was stimulated not only by the threat of the NRDA process being pursued by the Fish and Wildlife Service but also by the unexpected intervention of the U.S. EPA.

In June 1997, the EPA announced that it was unhappy with the pace of development of a comprehensive cleanup plan for the Fox River and that it was seeking the state's support for placement of the river on the Superfund National Priorities List (NPL). The agency was aware of the progress that had been made by various technical work groups and the contributions of the Fox River Coalition, among other activities. But the EPA was still dissatisfied, and it believed Superfund proceedings could speed up the cleanup process. In public meetings, agency staff highlighted the need to act more quickly by noting that the lower Fox received a "very high score" on a preliminary assessment of risk to public health and the environment. That put it on a par with PCB-related Superfund sites on the Hudson River in New York, the Kalamazoo River in Michigan, and the Sheboygan River in Wisconsin (Campbell 1997b; U.S. EPA 1997).

The EPA's argument was not persuasive to many in the region. Nearly all state and local government officials and representative of the paper mills denounced its action. Paper mill officials asserted that a Superfund designation could "stigmatize" the community and lead to further conflict and delays. They continued to praise a "public-private partnership" as the model to follow. WDNR's Meyer, agreed, vowing to press on with his preferred voluntary cleanup agreement with the mills (Hildebrand and Campbell 1997; Campbell 1997c). In sharp contrast, area environmentalists applauded the EPA decision, saying that listing the site would hasten cleanup of the river and force the polluters to pay.

Within three weeks of the EPA announcement, federal, state, and tribal officials, who had been negotiating for about eight months, were able to finalize some agreements. The U.S. Fish and Wildlife Service, EPA, National Oceanic and Atmospheric Administration, WDNR, and Oneida and Menominee Indian tribes consolidated their efforts and pledged to work together through a "consensus committee": the Lower Fox River Intergovernmental Partnership. The new approach replaced the voluntary cleanup effort that the WDNR had been working on with the paper mills since 1992, under the auspices of the Fox River Coalition. It also put the FWS's damage assessment initiative and the EPA's Superfund investigation into the background temporarily. As of spring 1999, those two federal assessments were still under way, and the agencies were considering using their regulatory authority if negotiations could not produce a satisfactory, voluntary cleanup agreement with the mills.

In July 1998, the EPA formally proposed listing of the lower Fox River on the Superfund's NPL, arguing that "the site poses very serious human health and ecological risks which are not being adequately addressed"(Campbell 1998). EPA officials noted that this was the first time a site had been proposed for Superfund designation without a governor's concurrence. As the agency began accepting public comments on the draft proposal, its action set off another round of intense debates in the region over the effect such a decision could have on the area's economy, and the advantages and disadvantages of a federally led cleanup effort. For its part, the EPA maintained that Superfund listing in no way precluded a continuation of ongoing negotiations over a voluntary cleanup plan between the state and the paper mills. Such an agreement could be reached at any time, officials argued, with the state, rather than the EPA, playing the lead role.

In late 1998, estimates for cleanup costs ranged from about $600 million to over $2 billion, depending on the extent of the effort and the technologies that would be used. By March 1999, however, the DNR released a new risk assessment and economic analysis that lowered the estimated costs for "feasible" cleanup alternatives for the river alone (excluding the bay) to between $150 million and $728 million (WDNR 1999). Despite the new studies, by April 1999, disagreement continued over PCB risks and cleanup options. The mills wanted to defer large-scale sediment dredging until they could complete their pilot projects and get a better sense of costs and the efficacy of removal technologies such as hydraulic dredging. They also argued that leaving the contaminated sediments in place might be a viable alternative if they posed no direct threats to human health.

The consensus of area scientists and agency technical staff, however, was that removal of nearly all of the PCBs from the lower Fox was essential for both human and ecological health. Yet even they agreed that large-scale cleanups might not produce the desired reduction in PCB levels in the river and bay. Thus they concluded that cleanup should proceed in stages and be rigorously evaluated throughout the process for the efficacy of remediation methods, with adjustments in cleanup strategies as needed. The group argued further that some portion of the funds made available by a settlement should be used for ecological repairs, such as habitat restoration, and prevention of nonpoint source pollution. They believed that such expenditures would be a more cost-effective way to reduce risks to ecosystem health in the river and bay (Barker and Kennedy 1998). Table 4.1 summarizes key developments that have shaped local water quality actions over the past several decades.

Conclusions

The Fox-Wolf Basin case illustrates the challenges that most communities and regions face as they attempt to move toward the long-term goal of sustainability. The 1970s-era environmental protection policies have accomplished much, as control of point sources and improved water quality in the Fox River Basin indicate. Yet further progress requires adjustment in those policy strategies to enhance society's capacity to deal with the more intractable problems of nonpoint sources such as the river sediments contaminated with PCBs and runoff from agricultural lands. The high cost and economic impacts on communities of cleanup actions propel the search for cost-effective means and for ways to encourage cooperation between the public and private sectors. Such cooperation can also help to build the mutual trust and consensus that are essential for creating sustainable communities.

The case reported on in this chapter traces the evolution of water quality issues across two of the epochs introduced in chapter 1, and the beginning of the third. It highlights the important efforts at collaborative decision making about water quality issues in Northeastern Wisconsin, a state widely considered to be among the most environmentally progressive in the nation. The work of the Remedial Action Plan and the Fox River Coalition is a story of both success and failure. Public-private partnerships were formed, and they helped to establish effective working relations among business officials, state and local policy makers, scientists, environmentalists, and community leaders. Regional water quality problems were thoroughly investigated by the Green Bay RAP and the

Table 4.1
Chronology of the Fox River cleanup

1927	One of the earliest records of a public dispute about water quality in the Fox River Valley; city committee members observe poor water quality during boat trip.
1931	Green Bay Metropolitan Sewerage District (GBMSD) is formed after petition campaign. City's first sewage treatment plant is built.
1937	Bay Beach, a local swimming beach, is closed by state Board of Health. Final closure occurs in 1943.
1948	Newly formed, area Izaak Walton League calls for state action on water pollution.
1949	State orders installation of wastewater treatment facilities by Fox River municipalities and paper mills by 1951.
1957	Paper mills begin discharging PCBs into Fox River through recycling of carbonless copy paper. PCBs continue to be discharged through recycling.
1971	Paper mills voluntarily stop using PCBs because of increasing environmental concerns.
1972	Congress enacts Clean Water Act, requiring industries and municipal sewage treatment plants to meet water quality discharge standards. Court challenge requiring GBMSD to upgrade its treatment plant succeeds.
1976	Manufacture of PCBs is banned by the U.S. EPA under authority of the Clean Water Act. Restrictions on transport and disposal of PCBs are instituted as well in the mid-1970s.
1987	Industries and municipalities report that Fox River dischargers have invested more than $300 million in water pollution controls. Water quality has greatly improved because of limits on point sources.
1988	Green Bay Remedial Action Plan (RAP) is completed, identifying PCBs in Fox River sediments as most important contamination problem. RAP is adopted by state as part of its Water Quality Management Plan.
1989–1991	State Department of Natural Resources (DNR) and U.S. Fish and Wildlife Service (FWS) discuss a joint Natural Resource Damage Assessment (NRDA). State does not agree to begin it.
1992	Wisconsin DNR, paper mills, and municipalities form Fox River Coalition (FRC) as a voluntary river cleanup effort. Northeast Wisconsin Waters of Tomorrow (NEWWT) is formed and later becomes Fox-Wolf Basin 2000.

1994	The EPA's Mass Balance study links PCB releases to natural resource damage in the Fox River and Green Bay. FWS launches its NRDA, identifying seven paper mills (the Fox River Group) as potentially responsible parties.
December 1996	FWS notifies the state and the paper mills that it may file a formal notice of its intent to file suit for cleanup costs after a deadline of January 31, 1997.
January 1997	WDNR and paper mills agree to spend up to $10 million as a "down payment" toward Fox River cleanup, to fund a demonstration project and other studies. FWS issues formal Notice of Intent to sue the mills.
June 1997	U.S. EPA, citing health risks posed by PCBs, announces its intent to place a 39-mile segment of the Fox River and a section of the bay of Green Bay on its Superfund National Priorities List.
July 1997	Federal, state, and tribal officials, acting as the Lower Fox River Intergovernmental Partnership, sign an agreement to cooperate in devising a comprehensive cleanup plan for the Fox River.
July 1998	U.S. EPA formally proposes Superfund NPL designation for the lower Fox River. The agency receives a record number of public comments on the proposal (about 7,000), more than two-thirds of which favor listing.
1999	The WDNR and EPA expect to choose a final cleanup plan for the lower Fox River following additional public comment on scientific studies, and the EPA will make a final decision on the Superfund listing. Results of the NRDA to be released.

Sources: "Special River Renewal Report," *Green Bay Press Gazette.* September 14, 1997, p. A8, and authors.

Mass Balance Study, and, using collaborative and consensual approaches, action agendas were developed to restore damaged ecosystems and protect public health.

Yet cooperation and voluntary cleanup programs, especially those directed at contaminated sediments, could go only so far in light of the high costs of remediation, ill-defined goals, and limited public understanding of the issues. Scientific knowledge by itself could not drive the cleanup process. Support by policy makers, the public, and industry was crucial for success. Without it, neither the RAP process nor the FRC could achieve its goals. After five years of extensive discussion and negotiation, the FRC's public-private partnership resulted in only a modest tangible commitment to cleaning up the river—primarily the agreement to provide $10 million to initiate remediation at two demonstration sites. The major constraints that so limited this process seemed to be a reluctance to pay the high price of cleanup and continuing doubts that the anticipated ecological and health benefits could justify doing so. Under these conditions, the threat of intervention by the EPA and FWS acting under the authority of Superfund helped to spur negotiations over a final cleanup agreement that may lead to some degree of restoration of the area's once pristine waters.

The inability of collaborative and consensual decision making to fully replace conventional command-and-control approaches does not imply, of course, that it has little value. The years of cooperation between business and government, and other stakeholders in the region, helped to build mutual understanding and trust that will be vital to achieving community sustainability in the future. Equally important is the contribution that such collaboration makes to building public trust and confidence in regional water quality decision making. Eventually, some difficult choices will have to be made about how far to go in removing contaminated sediments, and in controlling other nonpoint sources of pollution such as agricultural runoff. How clean should the river be? And what level of residual risk, especially from toxic chemicals, will be acceptable to the public? Making such decisions is never easy, but their success in improving regional water quality is likely to depend critically on the public's understanding of the issues and its belief that the decision-making process was legitimate and open to community participation.

Thus even as the goals of epoch one and two remain incompletely realized, leading-edge thinking and action are moving toward the sustainability paradigm of epoch three. Indeed, by mid-1997, a multiplicity of efforts was under way to build sustainability in the Green Bay metropolitan area. Local policy makers, citizen groups, and business leaders increasingly began using the language of sustainability. Some of the smaller communities in the region were holding envisioning conferences to help define their long-term goals. And the Bay Area Community Council, a broadly based seven-year-old citizens group, sponsored a well-attended conference in April 1997 on how sustainable communities can be fostered. In 1998, the WDNR began a follow-up program, the Sustainable

Green Bay Initiative, using EPA pollution prevention grant funds. With the assistance of consultants who had worked on such initiatives across the nation, the project team planned two working conferences for 1999, which were to bring community leaders together to help define an overall vision for a sustainable Green Bay, and a series of strategic actions that could help to realize it.

The outcomes in this case suggest that collaboration among community stakeholders works best when diverse parties are seeking agreement on general cleanup goals—particularly as they deal with the inherent uncertainty of environmental and health risks and the anticipated costs of cleanup. This approach to decision making may be less suitable when consensus cannot be easily achieved on specific cleanup schedules, the means to be used, and the allocation of costs—and where participants believe they will benefit from deferring decisions. Under those conditions, the continuing presence of stringent federal environmental laws and standards, such as those embodied in the Clean Water Act and Superfund, complements and reinforces a collaborative decision-making process. In the case of the Fox-Wolf Basin, federal intervention promised to end what had become protracted discussion and delay, and it created a new opportunity for negotiation and agreement on the cleanup of the river.

This experience suggests that hybrid policies and planning processes may be needed to spur community sustainability in epoch three of the modern environmental era. We could, for example, combine the best aspects of a regulatory framework with market-based approaches, public-private collaboration, and public education on the issues. Communities could be given sufficient time to resolve local problems on their own through cooperative processes before more formal regulatory requirements, from the federal or state government, take effect. The challenge for each community, region, and state is to design and implement its own distinctive blend of policies and plans that best promises to inspire, support, and nourish a transition toward sustainability.

Notes

1. Starting in 1995, EPA's Water Office began a more determined effort to make the 305(b) reports from the states more uniform and meaningful for assessing water quality conditions across the nation. In addition, data compiled by the U.S. Geological Survey on stream water quality offer another source of information

on national trends during the 1970s and 1980s. These data support the EPA's conclusions about declines in pollution from point sources.

2. Some policy theorists, such as Deborah Stone (1996) would call this kind of approach a classic "powers" strategy. It seeks to change the way decisions are made in the hope of getting different (and "better") decisions. Others, such as Schneider and Ingram (1997), remind us that successful policies depend upon good analysis and design. They must be based on a thorough understanding of their target populations, such as industrial and municipal polluters, and their motivation to comply with regulatory mandates (typically, concern for the sanctions imposed for noncompliance), or on some expectations for cooperative behavior that will be encouraged by a process of collaboration with stakeholders.

3. As one sign of the breath of representation, the Citizen Advisory Committee of the initial RAP was comprised of representatives of business, industry, environmental groups, sports and boating organizations, agriculture, shoreline residents, and local government agencies. Many other groups were involved with the technical advisory committees as well.

4. This last goal reflects the operation of the port of Green Bay, which necessitates dredging a shipping channel, and thus disposal of dredge material that is contaminated to some extent. The manner in which it is handled affects water quality.

5. The Mass Balance Study involved investigators from the U.S. EPA, U.S. Geological Service, National Oceanic and Atmospheric Administration, WDNR, and a host of universities. Initiated in 1986, the study sought to accurately model contaminant transport through multiple media using principles of mass conservation. The idea is to determine the quantities of contaminants entering the Green Bay system; the quantities stored, transformed, or degraded within the system; and the remainder that leave the system. Researchers tracked four key contaminants, including PCBs, in the sediment, water column, and air of the Fox River and Green Bay system. The data collected are the most comprehensive available for toxic pollutants entering Lake Michigan, and eventually will allow policy makers to better understand how their actions can affect water quality. The Green Bay study was the first in the world to determine the presence, transport, and fate of bioaccumulating toxic substances in a river and bay environment (Mercurio, 1995, ii).

6. The Fish and Wildlife Service, located within the Department of the Interior (DOI), is authorized under CERCLA to conduct an NRDA when contamination exists, and to seek funds from responsible parties for cleanup. DOI's NRDA procedures were upheld by the federal courts in 1994, allowing the FWS to use such assessments. The FWS serves under the law as a "trustee" for migratory birds and waterfowl, endangered species, and migratory fish, giving it authority to act in the Fox River case.

References

Adler, Robert W. 1993. "Water Resources: Revitalizing the Clean Water Act." *Environment* 35 (November):4–5, 40.

Alesch, Daniel J. 1997. "New Strategies for Environmental Problems in Wisconsin: Breaking Out of the Box." *Wisconsin Policy Institute Report* 10, 2 (February): 1–26.

Barker, Nancy, and John Kennedy. 1998. Letter to Secretary George E. Meyer, Wisconsin Department of Natural Resources, from Nancy Barker, chair, Public Advisory Committee, and John Kennedy, chair, Science and Technical Advisory Committee, Lower Green Bay and Fox River RAP, April 30.

Beyer, W. Nelson, Gary H. Heinz, and Amy W. Redmon-Norwood, eds. 1996. *Environmental Contaminants in Wildlife: Interpreting Tissue Concentrations.* Boca Raton, FL: Lewis Publishers.

Bosley, T. R. 1978. "Loss of Wetlands on the West Shore of Green Bay." *Wisconsin Academy of Sciences, Arts and Letters* 66:235–245.

Botts, Lee, and Paul Muldoon. 1997. *The Great Lakes Water Quality Agreement: Its Past Successes and Uncertain Future.* Hanover, New Hampshire: Institute on International Environmental Governance, March.

Campbell, Susan. 1998. "Fox Proposed for Superfund: Potential Remains for Voluntary Cleanup." *Green Bay Press-Gazette,* July 10, A1–2.

Campbell, Susan. 1997a. "Bickering Threatens to Delay Fox Cleanup." *Green Bay Press Gazette,* May 14, A1–2.

Campbell, Susan. 1997b. "Fox 'High on List' for Superfund." *Green Bay Press Gazette,* June 20, A1–2.

Campbell, Susan. 1997c. "DNR, Feds on 'Same Side of the Table'." *Green Bay Press-Gazette,* July 12, A1–2.

Campbell, Susan. 1997d. "Future of Fox River at Stake." *Green Bay Press-Gazette,* September 14, 1.

Cohen, Michael, James March, and Johan Olsen. 1972. "A Garbage Can Model of Organizational Choice." *Administrative Science Quarterly* 17 (March):1–25.

Colborn, Theodora E., Alex Davidson, Sharon N. Green, R. A. (Tom) Hodge, C. Ian Jackson, and Richard A. Liroff. 1990. *Great Lakes, Great Legacy?* Washington, DC: Conservation Foundation.

Council on Environmental Quality. 1997. *Environmental Quality: Twenty-fifth Anniversary Report.* Washington, DC: CEQ.

Davies, J. Clarence, and Jan Mazurek. 1998. *Pollution Control in the United States: Evaluating the System.* Washington, DC: Resources for the Future.

Freeman, A. Myrick III. 1990. "Water Pollution Policy." In Paul R. Portney, ed. *Public Policies for Environmental Protection.* Washington, DC: Resources for the Future. Pp. 97–149.

Harris, Hallett J. 1994. *The State of the Bay, 1993: A Watershed Perspective.* Green Bay, WI: University of Wisconsin-Green Bay Institute for Land and Water Studies, August.

Harris, Hallett J., Robert B. Wenger, Victoria A. Harris, and David S. Devault. 1994. "A Method for Assessing Environmental Risk: A Case Study of Green Bay, Lake Michigan." *Environmental Management* 18, 2:295–306.

Harris, Hallett J., Daniel R. Talhelm, John J. Magnuson, and Anne M. Forbes. 1982. "Green Bay in the Future—A Rehabilitative Prospectus." Ann Arbor, MI: Great Lakes Fishery Commission, September.

Harris, Hallett J., and Denise Scheberle. 1998. "Ode to the Miner's Canary: The Search for Effective Indicators." In *Environmental Program Evaluation: A Primer*, ed. Gerrit J. Knaap and Tschangho John Kim. Urbana, IL: University of Illinois Press. Pp. 176–200.

Harris, Victoria A. 1992. "From Plan to Action: The Green Bay Experience." In *Under Raps: Toward Grassroots Ecological Democracy in the Great Lakes Basin*, ed. John H. Hartig and Michael A. Zarull. Ann Arbor: University of Michigan Press. Pp. 37–58.

Hildebrand, Scott, and Susan Campbell. 1997. "EPA: Add Fox to Superfund List." *Green Bay Press-Gazette*, June 19, A1–2.

Johnson, Bruce N. 1996. "A Case Study of the Fox River Coalition: An 'Innovative' Approach to Contaminated Sediment Remediation in the Lower Fox River." De Kalb, IL: Department of Political Science, Northern Illinois University, unpublished paper.

Knaap, Gerrit J., and Tschangho John Kim, eds. 1998. *Environmental Program Evaluation: A Primer*. Champaign, IL: University of Illinois Press.

Knopman, Debra S, and Richard A. Smith. 1993. "Twenty Years of the Clean Water Act." *Environment* 35 (January–February):17–20, 34–41.

Kraft, Michael E. 1996. *Environmental Policy and Politics: Toward the Twenty-First Century*. New York: HarperCollins.

Mercurio, Jo. 1995. "The Fox River Coalition: A Regional Partnership Dedicated to Cleaning up Contaminated Sediment and Improving Water Quality in the Fox Valley." Madison: Department of Natural Resources, WR-382-95.

National Academy of Public Administration. 1995. *Setting Priorities, Getting Results: A New Direction for EPA*. Washington, DC: NAPA.

Ringquist, Evan J. 1993. *Environmental Protection at the State Level: Politics and Progress in Controlling Pollution*. Armonk, NY: M. E. Sharpe.

Schneider, Anne Larason, and Helen Ingram. 1997. *Policy Design for Democracy*. Lawrence, KS: University Press of Kansas.

Smith, Peyton L., Robert A. Ragotzkie, Anders W. Andren, and Hallett J. Harris. 1988. "Estuary Rehabilitation: The Green Bay Story." *Oceanus* 31, 3:12–20.

Stone, Deborah. 1997. *Policy Paradox; The Art of Political Decision Making*. N.Y.: W. W. Norton.

Sullivan, John R., and Joseph J. Delfino. 1982. "A Select Inventory of Chemicals Used in Wisconsin's Lower Fox River Basin." Madison, WI: University of Wisconsin, Sea Grant Institute.

U.S. Environmental Protection Agency. 1998. *National Water Quality Inventory: 1996 Report to Congress*. Washington, DC: Office of Water, April.

U.S. Environmental Protection Agency. 1997. "U.S. EPA's Superfund Role: Lower Fox River Cleanup." Chicago: Office of Public Affairs, Region 5, July.

U.S. General Accounting Office. 1997a. "Environmental Protection: Challenges Facing EPA's Efforts to Reinvent Environmental Regulation." Washington, DC: GAO/RCED-97-155, July.

U.S. General Accounting Office. 1997b. "Regulatory Reinvention: EPA's Common Sense Initiative Needs an Improved Operating Framework and Progress Measures." Washington, DC: GAO/RCED-97-164, July.

Weber, Edward P. 1998a. *Pluralism by the Rules: Conflict and Cooperation in Environmental Regulation.* Washington, DC: Georgetown University Press.

Weber, Edward P. 1998b. "Successful Collaboration: Negotiating Effective Regulations." *Environment* 40 (November):10–15, 32–37.

White, David, Baumgart, Paul, and Bruce Johnson, eds. 1995. "Toward a Cost-Effective Approach to Water Resource Management in the Fox-Wolf River Basin: A First Cut Analysis." Green Bay: Northeast Wisconsin Waters of Tomorrow.

Wisconsin Department of Natural Resources. 1999. *Draft Remedial Investigation, Risk Assessment, and Feasibility Study for the Lower Fox River.* Madison: WDNR, March.

Wisconsin Department of Natural Resources. 1988. *Lower Green Bay Remedial Action Plan for the Lower Fox River and Lower Green Bay Area of Concern.* Madison, WI: WDNR, PUBL-WR-175-87 Rev 88, February.

Wisconsin Department of Natural Resources. 1993. *Lower Green Bay Remedial Action Plan 1993 Update for the Lower Green Bay and Fox Area of Concern.* Madison: WDNR, September.

5
Local Open-Space Preservation in California

Daniel Press

A central premise of sustainability is that good land, in sufficient amount and of suitable quality, needs to be passed down from generation to generation. Good land is used for many purposes—agriculture, industry, housing, transportation, recreation, and habitat—and more destructive and irreversible uses clearly prevent future generations from putting land to its optimal, long-run use.

Communities at the municipal and county level exercise tremendous influence on land use within their jurisdictions, and thus one important dimension of any community's sustainability is its ability to preserve land in agriculture, open space, and wildlife habitat, which, in effect, helps preserve long-run options and uses. Also, green spaces, parks, forests, and farms provide an immediate connection to the places where people live. Open space also has an aesthetic appeal. Thus land preservation is one of the most enduring and rewarding commitments a community can make to future generations.

Looking back on how and why some communities succeed in preserving open space provides valuable insight into the opportunities and constraints faced in the movement toward sustainability. Scholars, social critics, and activists interested in the sustainability of communities can benefit from systematically identifying those communities that are in some way implementing sustainable land-use practices. In this chapter, I examine California, which has served as a leader in many aspects of environmentalism, such as in air, water, and toxic pollution, for a particularly illuminating case of land preservation as an indicator of progress toward sustainability.

California is well known for its extensive federal and state wilderness areas, parks, and open spaces. The role local actors play in land preservation is much less commonly understood, however. This chapter first reviews the patterns of local open-space preservation, noting especially how preservation practices have changed in the post-World War II period. Much of this change has mirrored the evolution in environmental policy outlined in chapter 1. What quickly becomes clear is that the patterns and practices of local preservation vary widely from county to county such that no single explanation adequately accounts for all the observed variation.

Patterns of Local Preservation

To begin with, California has fifty-eight counties, eleven of which have 60 percent or more federal land and stand outside the "normal" pattern of local-based preservation. Data on park and open-space parcels was gathered for the remaining forty-seven counties. In gathering the data, parcels smaller than 10 acres were excluded on the assumption that many of these smaller holdings are made up of municipal sports and fairground facilities. Counting only lands that are wholly owned by cities, counties, and special districts, local actors have acquired approximately 585,000 acres since the 1920s. Adding in gifts to the state parks system that were largely acquired or brokered by local actors, and lands acquired by local land trusts, raises that figure to as much as 1.0 million acres. This figure may not appear very large relative to the state's total acreage (about 100 million acres), or federal land in the state (44.6 percent of the state's land area is in federal ownership, about 45 million acres[1]), but it is on a par with what the state Department of Parks and Recreation (DPR) has acquired and managed during this period (about 1.3 million acres). More important, local open-space preservation demonstrates substantial investment of local political, social, economic, and environmental resources.

Local open-space preservation in California is a widespread phenomenon that has occurred steadily throughout the twentieth century. Perfectly commensurable data over time are not available, but for rough comparison, Dana and Krueger (1958, 105) estimated that county and municipal parkland holdings in California amounted to approximately 260,000 acres in 1957. Almost twenty years later, Donald Ito (1977) inventoried local public landholders and found that regional parks (greater than 15 acres each) totaled 472,890 acres. A 1987 survey conducted by the state Department of Parks and Recreation (DPR 1988, 7) estimated that local governments held 568,348 acres of parks (including all sizes).

Cities, counties, special districts, and land trusts continue to successfully arrange for the purchase and transfer of substantial acreage. In the San Francisco Bay area and environs alone, the establishment of regional parks and open-space districts spans the century. The extensive holdings of the East Bay Regional Park District (in the San Francisco Bay area) were secured in Contra Costa and Alameda counties during the early 1930s. In 1934, at a time when the regional water district was divesting itself of watershed lands, county residents put on the ballot—and approved, 72 percent in favor—an initiative to buy expensive ridge-top lands through property taxes (Olmsted Brothers and Hall 1930, 1984). Rather strikingly, the vast majority of voters in two semi-urban counties voted to tax themselves—during the Depression—to preserve open space. Further south, in Santa Clara County (home of Silicon Valley), park and open-space acquisition has had a long history: the county parks department was created in 1956, the Midpeninsula Regional Open Space district was created in 1972 (with strong advisory approval from voters), and the Santa Clara Open Space Authority was created in 1994 (also with strong advisory approval from voters).

Land acquisition and protection continues to be a very active pursuit at the local level. For example, data on the nine counties of the San Francisco Bay area[2] show that they had set aside some 268,000 acres through local efforts as of 1998, with at least 50,000 acres of these—just about 19 percent—having been acquired since 1988 alone. Similarly, California's special districts have been consistently active. The 154 special districts entrusted with a parks and recreation mandate spent over $940 million (in 1995 dollars) in land acquisition costs alone between 1969 and 1995.[3] These expenditures vary from year to year, but averaging $30 million per year, almost half of the spending occurred since 1982.

Counties vary widely with respect to the amounts and types of open space protected by local efforts, from zero acres per thousand residents to over 200 acres per thousand residents. By most any measure, local open-space preservation efforts represent an impressive accomplishment.

However, the aggregate data mask the land-use struggles fought parcel by parcel, decade by decade. Usually, preservationists lose more than they gain if the history of development, and habitat, wetland, or agricultural land loss is any indication. The few successes reflect sustainability in action: local actors taking a long view of their resources, balancing social and biological needs, building capacity for environmental policy and action, and relating their local activities to broader national and international trends.

The Evolution of Local Open-Space Preservation Policy Patterns

The first half of the twentieth century saw slow, steady land acquisitions, primarily for recreation purposes, often with the state government providing major funding. As the first epoch of the environmental movement swept through California in the 1960s, local, state, and federal funds for local open-space preservation grew tremendously. Local governments exercised a new-found boldness to regulate on behalf of environmental interests, but this period was to be a short-lived—from about 1964 to 1978—ending with the passage of the antitax Proposition 13.[4] Epoch two strategies emerged out of necessity, beginning in the mid-1980s, when communities were compelled to greatly increase their reliance on public/private partnerships for conservation. Meanwhile, public and private actors alike have become more innovative at cultivating public support for open-space preservation.

The more recent changes need to be understood in the light of other major changes in the state. The four key ones are changes in the relationships between levels of government and between government and civil society, in the objectives of land preservationists, in the ability of local government to raise necessary funds for land preservation, and in the kinds of policy approaches and tools that local governments have available for preservation purposes.

Intergovernmental Relationships and the Role of Civil Society

The first of this list of changes reflects a shift in the predominant political and institutional arena for local open-space preservation decisions, which has moved from state and federal levels toward local and nongovernmental actors. Not surprisingly, as the leadership has shifted, there have been associated changes in the points of intervention, as well as in access to the policy process. Historically, local open-space preservation was pursued in more of a "top-down" fashion than it is today; elites (in and out of government, but generally in leadership positions) and technocrats were mainly responsible for initiating open-space preservation efforts, while implementation was left largely to regional and state managers. This is less the case today as the relationship between the local, regional, state, and federal governments have changed, as have the roles of nongovernmental organizations. The two most important ways in which these relationships have changed have to do with governmental authority and funding.

Governmental Authority   The state of California set the stage for many local open-space preservation efforts from the turn of the century through the early 1970s. It did so through three kinds of statutory thrusts, one designed to increase local powers of land acquisition, another to motivate voluntary conservation arrangements between landowners and local governments, and a third aimed at increasing local government planning capacity. To begin with, the state has gradually defined and expanded the scope of local powers for land preservation. Beginning in 1905, and extending well into the 1950s, the California state legislature passed a series of statutes expressly granting cities and counties (and later, special districts) powers they would need in order to acquire lands for park, recreation, and open-space purposes. These powers allowed local governments to assess property taxes, condemn lands, and form special parks, recreation, or open-space districts.[5]

Second, the state legislature sought to motivate voluntary conservation by private landowners through a series of statutes adopted over a twenty-year period. In 1955, the legislature attempted to decrease development pressure on agriculture in Santa Clara County by allowing farmers to choose to have their lands zoned exclusively for agriculture (the so-called Greenbelt Law of 1955). A number of laws with statewide applicability followed through 1969, including the Scenic Easement Deed Act of 1959, the California Land Conservation Act of 1965 (the Williamson Act), and the Open Space Easement Act of 1969, and its 1974 and 1977 amendments (Barrett and Livermore 1983). The thrust of all these efforts was to permit (or prod) local governments to assess private property at the lower rates associated with lands set aside for agricultural or open-space preservation, not at the wildly soaring values of many exurban parcels. Thus, for example, the Williamson Act encouraged farmers to "enroll" their land in a conservation program whereby owners agreed to keep their land in agricultural or open-space use for a minimum of ten years; in exchange, owners paid much lower property taxes. Assembly Bill 1150 (the Quimby Act of 1965) allowed local governments to require that developers set aside a portion of their subdivisions as parks or open space, or pay fees for parkland acquisition and maintenance.

Third, the state legislature helped cities, counties, and regions to develop the necessary expertise for land use (and by extension, conservation) planning. As far back as 1927, the state allowed local government to form planning commissions. But the most important planning requirement came as part of the new consciousness of the environmental movement, when, in 1970, the legislature required cities and counties to adopt "open space elements" in their general plans.[6]

The federal government has also played a role, traditionally as a funding source through two principal programs, the Land and Water Conservation Fund (LWCF) and the old Housing and Urban Development (HUD) grant program for open space near urban areas. Each program constituted major sources of funding for local governments in the 1960s and 1970s, but the HUD program ended in the mid-1970s and the LWCF was virtually unfunded in the 1990s.[7]

Nongovernmental Organizations (NGOs) Typical of epoch two policy patterns, NGOs today are much more important than in the past: they still serve traditional roles (attempting to sensitize the public to open-space preservation issues and raising money), but increasingly they serve as brokers of acquisitions, principal funders, and land managers. They also cultivate mass support for open-space preservation, especially on ballot propositions.[8]

In 1985, the Land Trust Exchange (later, the Land Trust Alliance, a national clearinghouse based in Washington, D.C.) counted 32 land trusts in California, with an annual operating budget of almost $1.3 million (Land Trust Exchange 1985). By 1994, that figure had risen to 119 land trusts (Land Trust Alliance 1995), whose work is complemented by some 400 Coordinated Resource Management and Planning programs

(CRMPs), and well over 600 nongovernmental environmental organizations. According to the 1998 National Directory of Conservation Land Trusts, land trusts in the state were responsible for protecting some 536,000 acres in California (Land Trust Alliance, 1998). Their annual operating budgets, as of summer 1994, amounted to at least $9 million, and their membership stood at approximately 173,300.[9]

Some NGOs are organized primarily to raise revenues for governmental entities; two notable examples include the Peninsula Open Space Trust (POST), which was established in large part to raise funds for, and donate lands to, the Midpeninsula Regional Open Space District (MROSD) just south of San Francisco, and the Anza-Borrego Foundation, established in 1967 to benefit the Anza-Borrego Desert State Park (in San Diego County). Since 1983 alone, the Anza-Borrego Foundation was responsible for adding 12,800 acres to the Anza-Borrego Desert State Park at a cost of over $2.5 million.

Funding   State financing of local open-space preservation tends to come from three sources: statewide bond acts placed on the ballot, programs and/or funds created by the legislature expressly for funding park acquisition (some of these dedicated to local assistance), and general fund expenditures. Each of these was in decline or moribund in the 1990s.

In 1928, voters in the state of California overwhelmingly approved (almost 74 percent in favor) Proposition 4, a bond act raising $6 million for park acquisition and maintenance. As early as 1938, the state legislature allocated 30 percent of tideland oil royalties to state park acquisition and development. That figure rose to 70 percent in 1943, but the allocation was frozen from 1947 to 1954, when the federal government impounded the royalties pending settlement (in favor of the states) of disputes over tidelands jurisdiction (USORRC 1962). As already mentioned, after the 1928 ballot proposition, there followed another nineteen statewide measures (mostly since the onset of the environmental movement, between 1962 and 1990), largely devoted to raising funds for parks and habitat lands acquisition. Of the twenty total measures thirteen passed and seven were defeated (in 1962, 1980, 1990, and 1994). Equally important, no new park bond with major funding has been passed in a decade, although California legislators came close to putting a $600 million coastal protection and park bond on the ballot in 1998. Failure

to approve park bonds does not necessarily signal changing voter priorities. For instance, electoral support for finance measures tracks closely
with confidence in the economy (see Bowler and Donovan 1994) and the
legislature's ability to overcome the required two-thirds super majority
for moving generally popular park bond acts to the ballot.

Furthermore, the state has not compensated for these failed ballot
initiatives out of the state's general fund—quite the contrary. The state's
general fund paid for 53 percent of the state Department of Parks and
Recreation's budget in fiscal year 1988–1990, but declined as a proportion of the budget to 27 percent in 1995–1996 (Governor's Budget 1996).
And the DPR's budget has declined steadily since the early 1980s, partly
because of the 1990–1992 recession and partly because state funding
priorities changed. For example, in its fiscal year 1993–1994, the state
of California shifted $2.6 million in property tax revenue from local
government to public school districts, thereby especially affecting counties and special districts; these local governments depend more on property taxes than do cities (DPR 1994, 87). Most important for local
open-space efforts, local assistance funds administered by the state DPR
dropped from a two-decade high point of $145 million in 1989–1990 to
$8.4 million in 1993–1994. These figures reflected the state's fiscal strug-
gles during the recession of the early 1990s; local assistance has crept
back up to about $30 million per year since 1994–1995.

At the same time, federal largesse has all but dried up, as mentioned
earlier. One of the most important sources of federal funding for parkland
acquisition has been the federal Land & Water Conservation Fund
(LWCF). This fund was created in 1964, and relies mostly on royalties
from oil and gas leases on the Outer Continental Shelf. Allocations to
California have declined precipitously since the early 1980s. The LWCF
took in about $19 billion between 1964 and 1996, but only $8 billion
dollars have been used, nationally, for park and habitat acquisition. The
all-time high was around 1977–1978, when Congress approved about
$800 million per year for parks and national preserves; since the early
1980s, Congress has spent $100 million to $300 million per year for
parks and preserves (Rogers 1996a). Between 1965 and 1992, the LWCF
funded 334 state-level projects in California with a value of about $100
million, and 944 local-level projects with a value of approximately $130
million (DPR 1994, 42–43). Although California received as much as $26

million from the fund in 1979, the state has been allocated less than $5 million per year since 1986 (the allocation in 1995 was only $1.6 million).

Congress has used the unappropriated balances in the LWCF for deficit reduction; however, the 105th Congress signaled a new willingness to restore LWCF funds. The Congress passed an Interior Department budget that included $699 million for the LWCF, although this was not quite the victory it seemed for local efforts since the entire sum was earmarked for maintenance of existing national parks and for partial acquisition of the Headwaters Forest in northern California.

New Policy Objectives Justify Local Open-Space Preservation

During the first environmental epoch, open-space preservation was pursued mostly for recreational purposes, with a few notable exceptions (e.g., federal wildlife refuges, portions of a few state parks). Open-space preservation today is still heavily focused on recreation, but habitat conservation, agricultural land preservation, and growth management are increasingly important components. For example, most of the current mission statements, general plans, and multiyear plans for park and recreation special districts have adopted a wildlife or habitat management component. Many cities and counties have used their land use authority to adopt "special area zones" to demarcate sensitive habitat, riparian ecosystem protections, greenbelts, or greenlines surrounding dense urban zones.

In the second and third environmental epochs (since the early 1980s), statewide ballot measures have had habitat and wildlife protection goals in them, often explicitly earmarking substantial funds for acquisition and management of particularly valuable ecosystems. This was the case especially with Proposition 117, passed in 1990, which required about $30 million in fund transfers for wildlife and habitat protection.[10]

The Antitax Crusade

No observer of local government in California can ignore the fiscal crises visited upon cities and counties in the wake of the 1978 Proposition 13. The predominant political and institutional context for public policy has changed from resource abundance and policy experimentation to stopgap management. Local government revenues plummeted, from about $10 billion just prior to Proposition 13 to approximately half that amount (DPR 1994, 87). Other antitax and expenditure initiatives followed Proposition 13. Proposition 4 (1979) placed a ceiling on both state and local government expenditures. Proposition 62 (1986) and Proposition 218 (1996) extended the two-thirds majority approval requirement of Proposition 13 to virtually all the types of assessments, fees, or taxes used by local governments.

From an electoral perspective, the antitax initiatives dramatically decreased the number of local measures with revenue-raising objectives that can pass. For example, between 1986 and 1998, 78 tax and bond acts for parks and open space were put on local ballots around the state. Of these, 48 (or 61.5 percent) received a better than 50 percent vote in favor, but because of Proposition 13's requirement for a super majority of two-thirds or more votes, only 13 (about 16 percent) of the measures actually went into effect.

New Policy Approaches and Tools: Nonacquisition Conservation Measures

Full-fee acquisition (i.e., owning parcels outright) is the principal form of local open-space preservation discussed thus far. While acquisition can be the surest way to remove a developable property from future threats, it is an expensive strategy. Currently, parks and open-space authorities routinely pay $5,000 to $50,000 per acre for new acquisitions (Rogers 1996b; Midpeninsula Regional Open Space District 1995; Greenbelt Alliance 1992). Many of the most desirable parcels are attractive precisely because they are close to rapidly expanding cities and towns.

Moreover, land acquisition and management costs have gone up tremendously, much faster than the growth of local parks and open-space agency budgets (Press, Doak, and Steinberg 1996; DPR 1988, 1994). As costs have soared, local governments and nongovernmental organizations have turned to other conservation options, including special area zoning, transferable development rights (TDRs, in which development options are spatially shifted), conservation easements (in which development rights are donated by landowners), and purchased development rights (Wright 1994). Although these methods of conservation tend to be less expensive than outright purchase, they have their own limitations. Special area zoning, for example, regulates development rights in order to promote a wide range of public interest goals, including sensitive or riparian habitat protection, agricultural land preservation, erosion control, and watershed management, to name only a few. It is the most controversial of the alternatives to outright purchase, relying as it does on the private sector to bear the costs of preservation. Furthermore, a change in city councils or boards of supervisors can quickly undo these protections.

TDR negotiations tend to go smoothly only if there are relatively abundant and uncontroversial sites to transfer development—which is not the case with the many TDRs involving sensitive habitats (usually wetlands). Conservation easements can offer attractive "income and estate tax deductions equal to the value of the development rights retired" (Wright 1994), but landowners are often reluctant to give up development rights in perpetuity. Purchased development rights can thus be quite expensive, to the point of negating the potential savings of buying the rights separate from the land. In the case of both conservation easements and purchased development rights, the transaction costs (time, energy, legal counsel) of negotiating agreements among landowners, conservation groups, and local governments can be quite high.

Nonetheless, the numerous land trusts that sprang up around the state during the 1980s, as epoch two thinking and strategies took hold, have provided precisely the sorts of resources needed to overcome the transaction costs of negotiating conservation easements and purchased development rights. One of the largest purchased development rights agreements in the nation was negotiated in 1996 by the Monterey County Board of Supervisors, the Big Sur Land Trust, and a landowner with a substantial coastal ranch. The Monterey County Board of Supervisors used $11.5 million of state bond act funds, left over from the 1988 Proposition 70, to purchase the development rights to 3,550 acres of the El Sur Ranch (Rogers 1996c).

Variation in Local Open-Space Preservation: Some Preliminary Explanations

There are major changes in the institutional arrangements and patterns of land protection across the three epochs of the environmental movement evidenced in California. At the same time, there is significant variability within each of the epochs, which requires further exploration.

Here the field of political science can help, with its rich literature on state and local policy variations in general (see Boyne 1985, 1992; Brace and Jewett 1995; for reviews of earlier studies; Sharpe and Newton 1984; Bowman and Kearney 1987; Rice and Sumberg 1995; Stonecash 1996), and state environmental policy in particular (see Lester and Lombard 1990, for reviews of earlier studies). Most of this literature compares variations in state and local administrative activities, policy guidelines, records of administrative enforcement, and campaign promises (Sharpe and Newton 1984; Ostrom, Tiebout, and Warren 1961; Lewis-Beck 1977; Boyne 1985). The few studies actually focusing on what happens on the ground with respect to environmental protection go beyond outputs, and stress broader socioeconomic and political factors. For example, Ringquist (1993) built an integrated model relying on political-economic characteristics (wealth, political ideology), group influence (organized interests), and political system characteristics (i.e., professional legislature) to examine variation in actual state air and water quality.[11] He found that state economic wealth is not a dominant factor in either air or water quality regulation, and that strong regulatory programs did make a difference in controlling air pollution (and water pollution, to a lesser extent).

An additional consideration in understanding differences in local behavior, quite relevant to land preservation, is "environmental policy capacity" (Press 1998). The policy capacity framework draws on socioeconomic, political, and cultural variables to explain a community's actions on behalf of environmental goals. A policy capacity lens reminds us that, like the fifty states, many communities in California have long-standing distinctive, local political cultures. What is politically possible in one county might be unthinkable in another, almost irrespective of the environmental epoch in which policy makers are operating. Communities manifest their different political cultures in many ways that are crucial to the outlook for sustainability. For example, some counties in California are more willing and able to identify and respond to environmental policy problems and opportunities. In some cases, a larger tax base permits more open-space acquisition; in others, residents have strong attachments to their landscapes, coming to expect (and demand) strong protections from their local governments.

Understanding why policy capacity and political culture vary is more important than ever in the current and coming environmental epochs. Today local and regional communities are more and more likely to be left fending for themselves as the state and federal governments retreat from their roles as top-down managers and funders. The examination later in this chapter of five potential explanations for this variation in local open-space preservation shows, first of all, that there is no simple way of understanding local land protection practices. Yet when all five are taken together they offer a plausible, interrelated set of reasons for the observed variation in local open-space preservation. And it is reasonable to think that these factors are also related to the success of a community's efforts to achieve sustainability, not just land preservation. The five factors are the wealth, urbanization, administrative capacity, environmentalism and place attachment, and civic engagement/social capital of a community.

Wealth

It is reasonable to assume that wealthier communities are more able and likely to spend scarce private resources and tax dollars on open space. Indeed, it makes sense that wealthier people energetically support land acquisition and zoning regulations to maintain the vistas that make their communities desirable. Similarly, wealthy communities might use open-space preservation as a way of gating their communities and buffering them from sprawl and lower-income development. Thus per capita income should be strongly correlated with locally protected open-space acreage.

Unfortunately, directly measuring this relationship is difficult, mainly because lands are acquired over several decades, but per capita income is usually measured only annually. The ideal way of measuring the relationship between wealth and acquisitions within a community would be to correlate both in a continuous time series. These data are not available, primarily because land acquisition dates more than five or ten years old are exceedingly difficult to reconstruct.

But economic resources are surely an important factor for any local government program. And it is possible to take a "snapshot" of per capita income and correlate it with locally acquired acres in a community for a

Table 5.1
Pearson correlation matrix

	URBAN	LINCOM	GOVREC	COAST	LOGPOP	LSTAT	LFED	VOTE	LAC
LINCOM	0.52***	1.000							
GOVREC	0.82***	0.49**	1.000						
COAST	0.33	0.54***	0.38*	1.000					
LOGPOP	0.82***	0.47**	0.99***	0.37*	1.000				
LSTAT	0.17	0.40*	0.34*	0.42**	0.34*	1.000			
LFED	−0.35	−0.46*	0.10	−0.10	−0.09	0.001	1.000		
VOTE	0.68***	0.80***	0.63***	0.33*	0.60***	0.27	−0.41**	1.000	
LAC	0.71***	0.59***	0.74***	0.44*	0.75***	0.41*	−0.07	0.63***	1.000

*   p < .01
**   p< .001
*** p < .0001

URBAN: Percent county population living in urbanized areas, 1990 census
LINCOM: Log county per capita income, 1995
GOVREC: Government (city plus county) receipts, 1995
COAST: Dummy variable, 1 = coastal county; 0 = inland county
LOGPOP: Log 1997 county population
LSTAT: Log 1997 California state park acreages, by county
LFED: Log federal acres by county
VOTE: Average percent approval on statewide park initiatives, by county
LAC: Log local-open space acreages

given year. (See table 5.1 provides summary correlation results applicable to the remainder of the chapter). What this shows is that per capita income is correlated with local open-space acres by county (and is significant in a regression analysis), although it leaves about two-thirds of the variance in acquisitions unexplained. The same is true for per capita income as a predictor of protected acreage per 1,000 residents.[12] Measuring wealth, alternatively, not on the basis of individual per capita, but at the community level as the per capita revenue raised by counties (city and county governments), is a bit more useful a predictor of acquisition trends.[13] In short, there is more to acquisitions than the simple wealth of the community.

Urbanization

The degree of urbanization is also associated with local open-space preservation activity; indeed, there is a reasonably strong correlation between the percent of a county population living in urbanized areas and the aggregate amount of local open-space acreages (see table 5.1). There are several ways in which the urbanization phenomenon contributes to local open-space preservation. First, urban sprawl is hard to ignore. A common plea heard around California is to prevent one's region from "becoming another Los Angeles." Development is particularly noticeable in the coastal foothill counties, where valleys are narrow enough to show dramatic land use changes. Park and open-space district managers frequently cite the visual impact of urban sprawl as a powerful motivator for growth management. Indeed, the enabling legislation of the Santa Clara Open Space Authority (1992) recommends that "priority for open-space acquisition should be focused on those lands closest, most accessible, and *visible* to the urban area"[14] (emphasis added).

Whether or not urban sprawl must be visible to most people in a community in order to effectively prompt support for open-space preservation is unclear. Certainly, many valley counties with flatter topographies preserve open space, although, typically, those areas surround rivers, lakes, or reservoirs.

A second, related, way in which urbanization encourages local open-space preservation is by prompting demand for recreation accessible to urban and suburban centers. As sprawl progresses, the average travel time to open space increases, particularly if mass transit serves only a few major corridors. For years, city and county planners, local legislators, and local parks officials have attempted to maintain open space access in ratios ranging from at least 2 acres per 1,000 residents to 15 acres per 1,000 residents (Nelson 1988; USORRRC 1962).

Finally, the construction and real estate industries play very important roles, certainly as the engines of development, but also as (willing or unwilling) partners in preservation. Developers know that land values increase with proximity to protected open space, and are thus sometimes amenable to voluntarily mitigating their projects by setting portions of their lands aside. Since the passage of the 1965 Quimby Act, California cities and counties have been authorized to pass ordinances requiring that developers either set aside land or pay fees for park dedications. These "development impact fees" can become quite substantial. Unfortunately, the fees are not reported to the state; they appear only in city and county budgets, and are not readily accessible. The surveys that have been conducted estimate that cities and counties in 1981 assessed an average of $800 per dwelling unit as park dedication fees[15] (Frank and Downing 1988). As of 1996, 268 cities and 30 counties had passed Quimby Act ordinances (OPR 1997). These jurisdictions issued permits for well over 50,000 new, privately owned housing units in 1994 (U.S. Department of Commerce 1995). If these 50,000 permits were each associated with, on average, $1,000 in park dedication fees, the total raised for acquisition and maintenance would be $50 million—more than the combined budgets of local land trusts, and on a par certainly with the local assistance grants currently being administered by the state Department of Parks and Recreation. Thus by the measure of park dedication fees alone, development and urbanization in California provide a substantial portion of the total public funds available for park and open-space acquisition and maintenance.

Urbanization is a complex bundle of social, political, and economic attributes. In some instances, urbanization may simply mask other demographic factors. For example, urban populations tend to support environmental causes much more than rural residents, and, indeed, urbanization is fairly well associated with support for environmental ballot initiatives (see table 5.1).

Administrative Capacity

Cities and counties in California also vary with respect to their administrative abilities. Some have large planning staffs with specialized expertise, others have access to sophisticated geographic information systems (GIS) providing valuable coverage of city and county land uses.[16] Not surprisingly, the larger jurisdictions have more planning staff. For example, tiny Alpine county (in 1993, a population of 1,200) reported in 1994 as having one planner for the entire county. The county of Los Angeles and its cities reported having 518 planners in 1994, or one planner for approximately every 17,000 residents (OPR 1995). Of course, the numbers alone do not indicate whether Alpine County can complete its planning tasks as well or better than Los Angeles. Differences in staff numbers are to be expected, but larger staffs might be better able to plan effectively for the particular lands and land uses involved, and may also permit specialization. For example, jurisdictions that have relatively large planning staffs can assign some grant-writing positions, and thereby take advantage of external (usually intergovernmental) funding sources.

Consider the case of competitive grants to the state Department of Parks and Recreation and the federal Land and Water Conservation Fund. Just under $80 million (unadjusted) was granted to cities, special districts, and counties for land acquisitions between 1964 and 1996. Jurisdictions competed for these funds; they were not block allocations. Larger jurisdictions certainly raised more funds than smaller ones; perhaps more interesting however, was that the number of planners per county was highly correlated with grants raised (r = 0.89), a much closer association than total population (r = 0.67).

Local agencies develop long-standing expertise that can be (but is certainly not always) marshaled in service of land preservation. For example, officials at the state Department of Parks and Recreation pointed out that they have come to expect timely, compelling, and comprehensive grant proposals from the East Bay Regional Parks District (EBRPD), which spans portions of Alameda and Contra Costa counties on the San Francisco Bay (Verardo 1997). And, indeed, the EBRPD had raised just under $7 million in competitive grants between 1964 and 1996, second only to the $8.5 million raised by the county of Los Angeles.

Environmentalism and Place Attachment

Support for environmental protection certainly varies from place to place in a large state like California. What is less clear is whether support remains constant over time, and, equally useful to know, whether and how environmental attitudes are "bundled." That is, do voters who are generally willing to approve bond issues for state parks also agree to impose restrictions and/or costs on industry for pollution control? Do county residents generally support environmental protection activities at the local level as much (or more) than state or federal actions? Survey data, were they available, would certainly help answer these questions.

California does have a long history of using the referendum and initiative process at both the state and local levels, from which we can infer answers to some of these questions. As mentioned previously, twenty measures concerning parks and open spaces have been on the state ballot since 1928, most of which were bond acts to finance acquisitions at the state and local levels. Table 5.2 lists the measures, the year in which they appeared on the ballot, bond funding levels, approval rate, and their stated purpose. Note that nineteen of the twenty measures were on the ballot between 1962 and 1994.

The returns by county are available for all of these ballot measures, and in each case a "yes" vote would be considered the proenvironment position. Statistical analysis of voting returns on these propositions shows that counties have voted consistently either above or below the state average.[17] In other words, counties that tended to vote above the state average for park bonds in the 1960s tended to do so again in the 1970s, 1980s, and 1990s—despite dramatic demographic shifts and changes in urban density throughout the state. Thus the local character of environmentalism, at least measured by these electoral returns, is persistent over time *and* varies from county to county.

Strong support at the ballot for statewide park measures also correlates reasonably well with local open-space preservation ($r = 0.63$, see table 5.1). This suggests that voters support land conservation at both the state and local levels, and that environmentalism is likely to influence community capacity to mount park acquisition and maintenance programs.

Local environmentalism is surely enhanced by the attachments people form to the regions in which they settle. Indeed, in implicit recognition of this, some activists respond to environmental degradation by calling for people to "reinhabit" and reconnect with their neighborhoods, cities, regions, and landscapes (Sale 1985; Etzioni 1995; Hempel 1995). Their presumption is that if citizens identify strongly with their place ("place" writ large, i.e., their community, landscapes, natural and cultural history), they will involve themselves energetically in the political and social choices affecting their towns and regions. Furthermore, in the face of negative pressures from urban sprawl, crime, poverty, and pollution, highly engaged citizens will often step up their sacrifices for, and commitments to, community renewal and environmental protection (Press 1994).

There is a substantial literature in several social science disciplines on the nature of emotional and cognitive "attachment" to place (Canter 1977; Altman and Low 1992; Mitchell et al. 1993). Most place theories "define place as a physical location with the following three components: the physical setting, human activities, and the human psychological processes relating to it" (Brandenburg and Carroll 1995). There is strong evidence in this literature that highly valued places are those in which meanings, values, and traditions become tightly bound to particular spaces through repeated activities and associations (Relph 1976).

Despite the attention given to "place," there are few empirical studies available to show whether strong attachment to place has distinctly political implications, in terms of substantially different policy outputs or outcomes. If place attachment does impact public policy, one would expect the clearest effects to show up in sustainable land-use practices at the community level, precisely because they constitute the policy arenas most amenable to local influence and control. Preliminary results from a survey we are conducting suggest that many Californians are, in fact, strongly tied to their communities. For example, of 4,100 California residents surveyed in 1997, 66 percent reported that they would be "sorry to leave" their communities, 79 percent expected to stay in their communities, and 73 percent agreed that feeling that they were "a part of their community" was important. Of those respondents who had moved from other towns or states, 55 percent said that the environmental quality of their new communities was better than the previous places in which they lived.[18] These early results do not yet permit regional or local

Table 5.2
Statewide park initiatives and bond acts, 1928–1994

Bond act	Year	Funding level, unadjusted dollars (in millions)	Statewide approval rate	Stated purposes of the act
Proposition 4	1928	6	73.7%	Land acquisition
Proposition 5	1962	150	47.3%	Land acquisition, recreation
Proposition 1	1964	150	62.4%	Land acquisition, recreation
Proposition 3	1966	0	55.5%	Authorizes legislature to define and establish assessment bases
Proposition 1	1974	250	59.9%	Land acquisition, recreation
Proposition 2	1976	280	51.5%	Land acquisition, recreation, and wildlife conservation (~5% of funds)
Proposition 3	1978	0	55.1%	Allows the state to sell surplus coastal property to state recreation and wildlife management agencies
Proposition 1	1980 (Primary)	495	47.0%	Land acquisition, recreation, and wildlife conservation
Proposition 1	1980 (General)	285	51.7%	Land acquisition, recreation
Proposition 2	1980	85	48.8%	Land acquisition in Lake Tahoe region, for recreation and habitat protection
Proposition 4	1982	85	52.9%	Land acquisition in Lake Tahoe region, for recreation and habitat protection

Proposition 18	1984	370	63.2%	Land acquisition, recreation, and wildlife conservation
Proposition 19	1984	85	64.0%	Fish and wildlife habitat acquisition
Proposition 43	1986	100	67.3%	Grants to local park agencies
Proposition 70	1988	776	65.1%	Wildlife, coastal, and parkland conservation
Proposition 117	1990	30 in fund transfers	52.4%	Wildlife and habitat protection
Proposition 128	1990	300	35.6%	Omnibus environmental protection statute
Proposition 130	1990	742	47.9%	Forest land acquisition
Proposition 149	1990	437	47.1%	Land acquisition, recreation
Proposition 180	1994	2,000	43.4%	Land acquisition, recreation, and wildlife conservation

comparisons, but they do suggest that the attachment phenomenon is important.

Civic Engagement and Social Capital

A final explanation for variations in local open-space preservation is that communities in California differ with respect to their levels of civic engagement and social capital, and that these two phenomena affect local land-use outcomes in important ways. Civic engagement refers to an ensemble of activities, including political participation, civic voluntarism, and membership in public interest associations, broadly defined (Putnam 1993, 1995; Rice and Sumberg 1995). Specifically, these activities include nonpolitical group membership, volunteering and charitable giving, voting, campaign work, campaign contributions, contacts with government officials, protests, informal community activity, attending local board meetings (or neighborhood associations), board or neighborhood association membership, affiliation with political organizations, and attending meetings of political organizations.

Social capital can be harder to define, but generally refers to a community's social networks and its norms of reciprocity and trust (Newton 1997). As Coleman put it:

Unlike other forms of capital, social capital inheres in the structure of relations between actors and among actors. It is not lodged either in the actors themselves or in physical implements of production. (Coleman 1988, S98)

Where social capital is high, individuals in a community can "do a good turn" with a reasonable expectation that their actions will be rewarded, if only by a stranger at some unspecified time in the future. Social capital reduces the small risks people take every day in the pursuit of their goals; it thus requires relatively high levels of generalized trust. Newton (1997, 577) points out that the social networks of individuals, groups, and organizations are also important elements of the social capital concept, ". . . because an ability to mobilize a wide range of personal social contacts is crucial to the effective functioning of social and political life."

Most important for our purposes, these two aspects of social capital, trust and networks, permit people to reach their collective goals and resolve their conflicts. If the civic engagement and social capital phenomena do influence a policy outcome like land preservation, communities must have ways of translating the generalized and latent pool of capital into action. In other words, it is not enough that communities be generically engaged, trusting, and networked; civic engagement and social capital may contribute to open-space preservation if communities actively marshal these civic resources toward preservation goals. Two factors can help effect these activities. First and foremost, community residents must value land preservation. As we noted earlier, there is a great deal of variation in voting on environmental ballot propositions; the proportion of California residents calling themselves environmentalists also varies around the state (Field Institute 1985). In essence, civic resources would constitute one element (among others, like economic resources) of a community's ability to protect land, while environmentalist values would constitute an element of its motivation or willingness to do so.

A second factor translating civic resources into land preservation would be a community's attitudes toward public goods. How does a community view the role of local government? Is land acquisition a legitimate, even desirable, pursuit of local government? Is taxation a desirable policy tool to achieve land preservation, or should citizens work through private organizations using private resources? Again, voting on statewide ballot initiatives helps us assess the variation on this factor. For example, California voters passed (by 56 percent) the latest of three antitax initiatives in November 1996 (Proposition 218). Only four counties failed to pass this initiative; not surprisingly, these include the three San Francisco Bay area counties of Alameda, Marin, and San Francisco. Residents in these counties tend to support relatively higher local taxes, demand high levels of public services, and protect relatively large acreages of local open space.

But are civic engagement and social capital decisive factors in explaining local open-space preservation in California? Most of the information scholars gather about such phenomena are based on survey data; but telephone surveys are not usually conducted to provide comparative results, at a substate level, for many counties or cities in a state. Once again, my survey results provide reason to believe that both civic engagement and social capital do vary across the state of California. Table 5.3 gives examples of county variation in response to some of the key civicness and trust questions in my survey.

These results are mirrored by a recent survey conducted by the Field Institute, which found regional differences in reported levels of social trust. For example, to the question, "generally speaking, would you say that most people can be trusted, or that you can't be too careful in dealing with people?," only 20 percent of Los Angeles county residents agreed that "most people can be trusted," while 41 percent of respondents in the nine-county San Francisco Bay area agreed. With respect to civic engagement, the survey found that 54 percent of San Francisco Bay area residents had worked with others in their own community to solve some community problem, while only 45 percent in Los Angeles had reported doing so.

Conclusion

Future survey results will help determine whether place attachment, civic engagement, and social capital correlate with local open-space preservation. At the broadest level, what is important is the emergence of the

Table 5.3
Sample survey results

Survey Question	Statewide mean (N = 4,100 respondents)	Range of county means (N = 27 counties)
Awareness of open-space issues in your community	39.0%, "yes"	23%, "yes" (Glenn County) to 70%, "yes" (Santa Barbara County)
Would you support a state tax bond measure for state parks acquisition, development, and maintenance?	75.7%, "yes"	61%, "yes" (Kern County) to 81%, "yes" (Santa Cruz County)
In the past year have you volunteered your time to any adopt-a-creek, -beach, -highway, or –park activities?	5.7%, "yes"	1.1%, "yes" (Glenn County) to 16%, "yes" (Humboldt County)
In the past year have you volunteered your time to any local government boards or commissions?	4.9%, "yes"	1.5%, "yes" (Los Angeles County) to 12%, "yes" (Amador and Merced Counties)
Have you ever attended a town meeting, public hearing, or public affairs discussion group?	53.0%, "yes"	42%, "yes" (Kern County) to 70%, "yes" (Santa Cruz County)
Frequency of voting in local elections	55.0%, "every time"	41%, "every time" (Riverside County) to 65%, "every time" (Amador County)
Generally speaking, would you say that most people can be trusted?	59.0%, "yes"	41%, "yes" (Riverside County) to 75%, "yes" (Humboldt County)

environmental movement and the evolution through at least two epochs so far. This has resulted in the shift from preservation as recreation in the preenvironmental era, to top-down, state-led acquisitions in the 1960s and 1970s, to the rise of nongovernmental organizations (NGOs), land trusts, and other local initiatives in the 1980s. In addition to the land trusts cited in the second part of this chapter, communities around the state have begun "adopt-a-creek" and "adopt-a-park" programs that mobilize literally thousands of volunteer hours on behalf of local streams and open spaces. These NGOs and volunteers raise funds, broker sensitive land acquisition deals, maintain and restore land and water resources, and cultivate popular support for local open space. They have become indispensable partners working on behalf of sustainability in their communities.

Local open-space preservation in California depends on a mixture of institutional, socioeconomic, and cultural/attitudinal factors. The precise mix of these factors that successfully facilitates preservation surely varies from community to community. How and why they change from place to place—and especially how they have changed over time—will be the subject of future research.

With respect to the future of local open-space preservation, the overall picture is that preservation has become more difficult to pay for, while state and federal governments have less of a direct role in acquisition. At the same time, more conservation groups exist (and a more diverse mix of NGOs), and they take on more of the open-space preservation tasks themselves. In essence, open-space preservation has become more democratic, but less secure. It is more democratic not only because of the proliferation of land trusts and environmental groups but also because there are so many NGOs and local governments capable of carrying out roles previously entrusted to just a few, centralized state and federal agencies (with a few regional exceptions). There is more redundancy in local open-space preservation, more willingness on the part of civil society to compensate for governmental retrenchment.

This horizontal spread of governance is only half the story, however. The number, funding, and membership of land trusts across the state is indeed impressive, but their efforts are one or two orders of magnitude smaller than what the state and federal governments can (and have) achieved (consider the cumulative annual budgets of all the land trusts— at about $9 million—versus that of the state Department of Parks and Recreation—about $200 million).

Future success will depend on innovative partnerships of the sort envisioned in epoch three as being pursued at all levels of government, and with active participation by NGOs and civil society. But many economic and political obstacles will remain. One valuable lesson from past efforts is that open space must become understood as a vital community asset, equal to a thriving downtown, a modern stadium, or safe, vibrant schools. Local actors working for sustainable communities would do well to track and publicize compelling indicators—the acres of parks and open space per 1,000 residents is a tried and true measure; the ratio of green space to "black space" (pavement) is also valuable. However, advocates of community sustainability will find their best asset in their region's land itself. The more a community's residents can experience the immediate and varied experience of open space, the more they will come to know and protect it, and seek to extend its reaches. For those willing to listen, a community's hills, woodlands, and streams tell us that sustainability, an idea so elusive in the abstract, is as certain as the open land at the edge of town.

Notes

1. By way of comparison, only two other states (Alasks and Nevada) have more acreage in federal ownership; only seven other states have a greater percentage of their land in federal ownership.

2. Marin, Sonoma, Napa, Solano, Contra Costa, Alameda, Santa Clara, San Mateo, and San Francisco.

3. State of California. State Controller's Office. *Report on Financial Transactions Concerning Special Districts*, various years, 1969–1995.

4. Proposition 13 was a constitutional amendment passed by initiative that limited property tax assessments to 1 percent (of assessed value) per year, and capped growth in property taxes at 2 percent per year (Fulton, 1991, 209).

5. See the California Public Resources Code, Sections 5000 to 6000, for an overview of most state regulations developing local government powers with respect to parkland acquisition and funding.

6. California Government Code, Section 65560.

7. For example, in 1972, the state Department of Parks and Recreation spent about $2 million in local assistance grants. The federal government spent over $8 million in Housing and Urban Development (HUD) grants in its undeveloped open-space lands program, and another $4.5 million from the Land and Water Conservation Fund. The sum total of land acquisition expenditures by all of California's special park and recreation districts in 1972 was a little over $7.8 million.

8. A handful of land trusts have been performing these roles for decades of course. For example, the Save-the-Redwoods League, founded in 1918, often matched state, local, or other private funds for acquisition, as did the Sempervirens fund, founded in 1900.

9. Not all land trusts reported how many acres they had protected and/or their annual budgets. Thus the actual, current figures are likely to be higher. Note also that the figures on acreage do not imply that all lands conserved by land trusts were protected by fee-simple acquisition. Many of these acreages are protected through conservation easements or other nonacquisition means.

10. In a related case, voters soundly (58 percent against) defeated Proposition 197 in 1996, a measure that would have permitted hunting of the state's mountain lions.

11. For example, to measure the air quality variable, Ringquist (1993) uses "percent change in state pollutant emissions, 1973–75 to 1985–87."

12. Both variables are non-normal. $R2 = 0.37$, linear regression with log transformed variables. F-ratio = 13.08, $p < 0.001$.

13. Residential tax base could also be used to explore the wealth thesis, since many special districts rely on property tax assessments for revenues. However, most city and county parks and open space agencies do not rely on property tax assessments. Thus I have not measured the residential property tax base against local acreage acquisitions.

14. CA Public Resources Code, Section 35152.

15. In 1981 dollars. The average would be about $1300 in 1995 dollars.

16. Geographic information systems permit users to computerize maps, and to overlay many different sorts of spatial data, such as the location of park parcels, roads, and creeks, with demographic, biological, and physical data. A GIS can permit users to rapidly and easily view spatial patterns of these data without redrawing paper maps.

17. G= 624, df = 57, $p < 0.001$. The chi-square test measures the difference between expected and observed frequencies. In this case, chi-square tells us that it is highly unlikely that the large differences we see in average county approval of park measures is due to chance.

18. In 1997, the Public Research Institute (PRI) at California State University, San Francisco, conducted a telephone survey of 4,100 California residents for

my Community and Conservation Project. For the survey funds, I acknowledge the generous support of the John Randolph Haynes and Dora Haynes Foundation and the U.S. Environmental Protection Agency (R 825226–01–0).

References

Altman, Irwin, and Setha M. Low, eds. 1992. *Place Attachment.* Human Behavior and Environment: Advances in Theory and Research, Vol. 12. New York, NY: Plenum Press.

Barrett, Thomas S., and Putnam Livermore. 1983. *The Conservation Easement in California.* Covelo, CA: Island Press.

Bowler, Shaun, and Todd Donovan. "Economic Conditions and Voting on Ballot Propositions." *American Politics Quarterly* 22 (January 1994):27–40.

Bowman, Ann O'M., and Richard Kearney. 1988. "Dimensions of State Government Capability." *Western Political Quarterly* 41:341–362.

Boyne, George A. 1992. "Local Government Structure and Performance: Lessons from America?" *Public Administration* 70:333–357.

Boyne, George A. 1985. "Review Article: Theory, Methodology and Results in Political Science—The Case of Output Studies." *British Journal of Political Science* 15: 473–515.

Brace, Paul, and Aubrey Jewett. 1995. "Field Essay: The State of State Politics Research." *Political Research Quarterly* 48:643–681.

Brandenburg, Andrea M., and Matthew S. Carroll. 1995. "Your Place or Mine?: The Effect of Place Creation on Environmental Values and Landscape Meanings." *Society and Natural Resources* 8, 5:381–398.

California Department of Parks and Recreation. 1994. *California Outdoor Recreation Plan, 1993.* Sacramento, CA: State of California Resources Agency.

California Department of Parks and Recreation. 1988. *Local Parks and Recreation Agencies in California: A 1987 Survey.* Sacramento, CA: State of California Resources Agency.

California. Governor's Budget, 1995–1996. Sacramento, CA. Governor, State of California.

California State Controller's Office. *Report on Financial Transactions Concerning Special Districts,* various years, 1969–1995.

Canter, David. Y. 1977. *The Psychology of Place.* London, UK: Architectural Press.

Coleman, James S. 1988. "Social Capital in the Creation of Human Capital." *American Journal of Sociology* 94 (Supplement):S95–S120.

Dana, Samuel T., and Myron Krueger. 1958. *California Lands: Ownership, Use, and Management.* Washington, DC: The American Forestry Association.

Etzioni, Amitai, ed. 1995. *Rights and the Common Good: The Communitarian Perspective.* New York, NY: St. Martin's Press.

Field Institute. 1985. *California Opinion Index: Environmental Issues.* San Francisco, CA: The Field Institute. Vol. 1, January.

Frank, James E., and Paul B. Downing. 1988. "Patterns of Impact Fees." In *Development Impact Fees: Policy Rationale, Theory, and Issues,* ed. Arthur C. Nelson. Chicago, IL: Planners Press.

Fulton, William. 1991. *Guide to California Planning.* Point Arena, CA: Solano Press Books.

Greenbelt Alliance, The. 1992. *The Bay Area's Public Lands: Findings from the 1992 Survey.* The Greenbelt Alliance, San Francisco, California.

Hempel, Lamont C. 1995. *Environmental Governance: The Global Challenge.* Covelo, CA: Island Press.

Ito, Donald H. 1977. *Recreation Technical and Information Paper 9: Recreational Acreage and Acres Per 1,000 Population for Cities, Counties, and Special Districts.* Sacramento, CA: State of California Resources Agency.

Land Trust Alliance. 1998. *1997 National Directory of Conservation Land Trusts.* Washington, DC: Land Trust Alliance.

Land Trust Alliance. 1995. *1995 National Directory of Conservation Land Trusts.* Washington, DC: Land Trust Alliance.

Land Trust Exchange. 1985. *1985–86 National Directory of Local and Regional Land Conservation Organizations.* Bar Harbor, ME: Land Trust Exchange.

Lewis-Beck, Michael S. 1977. "The Relative Importance of Socioeconomic and Political Variables for Public Policy." *American Political Science Review* 71:559–566.

Lester, James P., and Emmett N. Lombard. 1990. "The Comparative Analysis of State Environmental Policy." *Natural Resources Journal* 30:301–319.

Midpeninsula Regional Open Space District (MROSD). 1995. Purchase of 907-Acre Jacques Ridge Is Complete. Press release, January 13, 1995.

Mitchell, M.Y, Force, J.E., Carroll, M.S., and W.J. McLaughlin. 1993. "Forest Places of the Heart: Incorporating Special Spaces into Public Management." *Journal of Forestry* 9, 2:32–37.

Nelson, Arthur C., ed. 1988. *Development Impact Fees: Policy Rationale, Theory, and Issues.* Chicago, IL: Planners Press.

Newton, Kenneth. 1997. "Social Capital and Democracy." *American Behavioral Scientist* 40, 5:575–586.

Office of Planning and Research. 1997. *The California Planners' 1996 Book of Lists.* State of California, Governor's Office of Planning and Research, Sacramento, California.

Olmsted Brothers, and Ansel F. Hall. 1930, 1984. *Report on Proposed Park Reservations for East Bay Cities, California.* Prepared for the Bureau of Public Administration, University of California, reprinted 1984.

Ostrom, Vincent, Charles M. Tiebout, and Robert Warren. 1961. "The Organization of Government in Metropolitan Areas: A Theoretical Inquiry." *American Political Science Review* 55:835–842.

Press, Daniel, Doak, Daniel F., and Paul Steinberg. 1996. "The Role of Local Government in Rare Species Conservation." *Conservation Biology* 10, 6: (December):1538–1548.

Press, Daniel. 1998. "Local Environmental Policy Capacity: A Framework for Research." *Natural Resources Journal* 38 (Winter):29–52.

Press, Daniel. 1994. *Democratic Dilemmas in the Age of Ecology: Trees and Toxics in the American West.* Durham, NC: Duke University Press.

Putnam, Robert D. 1995. "Tuning In, Tuning Out: The Strange Disappearance of Social Capital in America." *PS: Political Science and Politics* XXVIII, 4 (December):664–683.

Putnam, Robert D. 1993. *Making Democracy Work: Civic Traditions in Modern Italy.* Princeton, NJ: Princeton University Press.

Relph, E. 1976. *Place and Placelessness.* London, UK: Pion.

Rice, T.W., and A.F. Sumberg. 1995. "Civic Culture and Government Performance in the American States." Unpublished manuscript.

Ringquist, Evan J. 1993. *Environmental Protection at the State Level: Politics and Progress in Controlling Pollution.* Armonk, NY: M.E. Sharpe.

Rogers, Paul. 1996a. "Park Piggy Bank Rescue Under Way," *San Jose Mercury News,* January 7, 1996, 1B, 4B.

Rogers, 1996b. "County Plans to Buy Big Area for Park Use," *San Jose Mercury News,* August 20, 1996, A1.

Rogers, 1996c. "Big Chunk of Big Sur Preserved," *San Jose Mercury News,* December 4, 1996, A1, A18.

Sale, Kirkpatrick. 1985. *Dwellers in the Land: The Bioregional Vision.* San Francisco, CA: Sierra Club.

Sharpe, L. J., and K. Newton. 1984. *Does Politics Matter? The Determinants of Public Policy.* Oxford, UK: Clarendon Press.

Stonecash, Jeffrey M. 1996. "The State Politics Literature: Moving Beyond Covariation Studies and Pursuing Politics." *Polity* 28, 4:559–579.

United States Department of Commerce, Economics and Statistics Administration. 1995. *New Residential Construction in Selected Metropolitan Areas.* Washington, DC: Superintendent of Documents.

United States Outdoor Recreation Resources Review Commission. 1962. *The Future of Outdoor Recreation in Metropolitan Regions of the United States.*

Washington, DC: The Outdoor Recreation Resources Review Commission, Vol. 3.

Verardo, Denzil. 1997. Deputy Director for Administration, California Department of Parks and Recreation. Personal communication, March 27.

Wright, J.B. 1994. "Designing and Applying Conservation Easements." *Journal of the American Planning Association* 60, 3:380–388.

III

Toward Community and Regional Sustainability: Leading Examples of the Transformation Process

6

The Challenge of the Environmental City: A Pittsburgh Case Study

Franklin Tugwell, Andrew S. McElwaine, and Michele Kanche Fetting

Around the industrial heartland of the United States, a number of cities have begun unusual environmental initiatives. By and large, these efforts involve no regulation, and little bureaucracy, and they avoid confrontation with economic actors. Instead, they focus on salvaging a ravaged landscape, rebuilding economies, and drawing people back into urban life. In Buffalo, the mayor is attempting to reclaim hundreds of acres that belonged to Republic Steel; the city plans parks, riverfront commercial establishments, and open space for the site. In Cleveland, aging infrastructure is giving way to new attractions. Racine, Wisconsin, has likewise established a sustainable development program, aimed at reinventing the onetime manufacturing center. Chattanooga, once described as a "dreary, depressing, industrial nowheresville," has declared itself to be the "Sustainable City." And Pittsburgh, the focus of this chapter, has launched an Environmental City Network aimed at building community support for its own first efforts to forge a new environmental ethos—and position itself to take advantage of economic opportunities as the nation struggles to respond to new environmental challenges.

Unlike the environmental confrontations of the past, these activities involve significant partnerships among public, private, and nonprofit actors. Rather than impeding market forces, they are seeking to channel them in ways that revitalize the urban core. Moreover, instead of imposing limits on economic activity, they are seeking to create new opportunity in ways that support environmental goals. At heart, many of them are attempts to discover local and cooperative means of moving communities toward a form of environmental renewal that will enhance quality of life while permitting a sounder economic basis for the future. In sum, many of these initiatives appear to signal the first signs of movement toward the "sustainable community" model (epoch three) presented in the first chapter of this volume. The form each takes of course, may be expected to depend heavily on the physical attributes of the locality involved as well as its distinctive history.

Pittsburgh's relationship with the environment has been viewed through many lenses. For much of the city's history, the natural world was viewed as an unlimited resource to be exploited at will. Beginning in the 1910s and culminating in the Pittsburgh Renaissance immediately following the Second World War, a period of bureaucratic control over the worst environmental excesses began that continues to this day. In the 1970s, the city and the region handed this responsibility off to the federal government. For most communities, this was something new, but Pittsburgh was already ahead of the game. Because of the region's early investment in air and water cleanup, the regulations posed little difficulty until the 1990s.

In the past three years, a somewhat different story has emerged. Federal and state command-and-control mandates for clean air compliance, sewage maintenance, and industrial site cleanup have imposed significant costs and engendered local opposition. At the same time, discrete pollution control regimes have been slowly giving way to a more comprehensive local approach to natural resource use and ecosystem management. It is a transition that is as compelling as it is difficult. Once an economy based on the extraction of natural resources, the processing of those resources, and the export of them to end users, the Pittsburgh region's economy is gradually developing into one that will require regional amenities and a high quality of life, in order to grow and attract new technology-based companies and high-performance manufacturers rather than old-line basic industries. In sum, Pittsburgh is in the early stages of an environmental reawakening.

While these efforts may come to represent significant changes, they are nevertheless still nascent and may yet give way to older confrontations and leave little behind beyond symbolic or cosmetic changes. Difficult issues of city governance, urban sprawl, pollution control, and job creation remain barriers; and as difficult trade-offs become apparent, conflict can threaten consensus. In Pittsburgh, as in many cities, commitments to goals associated with sustainable development—as understood in this book—are still highly contingent and are not widely shared by the influential players who determine the future of urban economies.

In Pittsburgh's case, the prospects for a deepening of the sustainable community initiative depend heavily on the ability of those involved to demonstrate that the initiative is compatible with rapid economic recovery. Local efforts to achieve sustainability should not weaken the city or region's ability to compete with others—in the United States or abroad—for job growth. Insofar as this remains a critical determinant, the local efforts will remain dependent on the success of national and global frameworks that reward local and regional initiative.

Judged by the analytic categories advanced in this volume, Pittsburgh as a whole is still ensconced in epochs one and two. Its principal leadership groups—city and county government and business elites—think of environmental issues mainly as the need to formulate responses to state and national demands for pollution control by regulation. These same leaders are often aware of the kinds of tools and approaches that characterize epochs two and three, but that awareness has not been accompanied by understanding, acceptance, or concrete attempts to fashion new policy instruments.

Within this broader framework, however, there has emerged a cluster of change agents determined to alter the state of affairs. Made up of environmental advocacy groups, foundation leaders, academics, and a growing network of small businesses, this cluster has begun the hard work of building a new environmental ethos—replete with collaborative agreements, new organizations, and informal networks. It is within this framework that Pittsburgh developments have begun to reflect epoch three thinking. Although there are only glimpses, and it is questionable whether the city as a whole is moving in that direction, the core environmental leadership responsible for these economic and environmental initiatives is promising. The pages that follow chronicle the historic accomplishments of pollution control in Pittsburgh—admirable by any measure—and set forth the achievements and plans of those seeking to move the city into a new era.

Hell with the Lid Off—and the Cleanup

Pittsburgh is located at the confluence of the Monongahela, Allegheny, and Ohio rivers, which carve out the hilly terrain and distinguish one of the most unique and lovely urban environments in the world. The city is the focal point of the Ohio River Basin—a region of 204,000 square miles, covering parts of fourteen states, and including a population of nearly 25 million (almost 10 percent of the U.S. population).

While the industrial revolution did not originate in Pittsburgh, in many respects the transformation of one of America's most beautiful landscapes into a nightmare world of pollution and land contamination epitomizes the historical American extreme of disregard for the environment and disrespect for the land and its long-term values. The despoilation of the green hills and valleys of Southwestern Pennsylvania took more than a century, but was thorough.

By 1803, iron making was established as Pittsburgh's most valuable industry—surpassing even whiskey production, the community's dominant enterprise in the eighteenth century. Access to natural resources and transportation fueled large-scale industrial activity that transformed the landscape of entire regions. In Pittsburgh, ore, timber, and coal in outlying areas and a network of waterways and highways made it a center of heavy industry. Industrial expansion drove its communities. Restrictions on the use of resources or the placement of factories—let alone disposal of waste—were unheard of. One of the largest foundries in antebellum America could be found at Smithfield Street and Fifth Avenue in the heart of Pittsburgh. By 1850, Pittsburgh was known for its sooty skies. By the end of the century, the city was known as "hell with the lid off" (Reiser 1951, ch. 3, 191–194, 213; Steffens 1906).

The industrial city of 1900 was a remarkable development. Pittsburgh was, perhaps, worse than most, but it was certainly not alone. Smoke from coal-burning factories, ships, steam locomotives, and homes enveloped the community. Visitors noted barren hillsides and speculated that nothing could grow under such conditions. Margaret Byington, viewing the city from Homestead early in the twentieth century, wrote: "The trees are dwarfed and the foliage withered by the fumes; the air is gray, and only from the top of the hills above the smoke is the sky blue" (Byington

1996, ch. 3). In 1907, 622 people died from typhoid fever, the result of contaminated drinking water, more than in any other city in the United States. The temperature of the massive Monongahela River was known to top 120 degrees Fahrenheit. Narrow streets were choked with traffic. The riverfronts were given over to steel mills, rail lines, parking lots, and barge traffic. A handful of parks were clustered in the city's wealthy east end communities, but playgrounds and public recreation were otherwise unavailable. Sanitary conditions in working-class neighborhoods were deplorable. Hundreds of workers were injured or killed in the city's factories. Sewage treatment was not available, and water lines reached working communities only intermittently. Workers' housing was a shambles. In short, life in industrial Pittsburgh was a Hobbesian existence—nasty, brutish, and short (Tarr 1996a, 181–183; Greenwald 1996, 133).

As the costs of industrialization became increasingly apparent, questions about Pittsburgh's development pattern grew more urgent. The city grew exponentially during the industrial revolution. From a city of 50,000 in 1860, it topped 300,000 in 1900, and became one of the largest cities in America. It was home to the nation's first oil refinery and its first steel mill. The city expanded its area from a little over 1 square mile to 30 square miles. After 1907, when the North Side was annexed, Pittsburgh approached a half-million residents. For more than one-hundred years, this growth had been conducted without regulation or management. The result was a mixture of industrial, commercial, and residential activities crammed into each region of the city and surrounding communities. Streets and essential services were likewise mismatched to the needs of the landscape, and there was no notion of land use controls (Tarr 1989a, 228, 240).

Early Environmentalism and City Planning

The rise of a professional and business elite late in the nineteenth century brought with it a new urban environmentalism that sought both voluntary controls and rational planning. Turn-of-the-century Pittsburgh spawned a Civic Club, an Engineers Society, a Chamber of Commerce, and other civic organizations devoted to urban improvement. These organizations sought an efficient city freed of the burdens of smoke, sewage, traffic, and dilapidation. In the case of the Civic Club, the effort went beyond aesthetics and was a moral issue: a dirty environment led to crime and immorality. By providing a healthy and attractive city with recreation and open space, many professionals believed that working-class pathologies could be ameliorated and the well-being of urban residents improved.

Opportunities for public-sector partnerships, however, were limited. The Flinn-Magee Republican machine had ruled the city since 1879, and did not embrace reform with any enthusiasm. After 1906, however, Democratic Mayor George Guthrie opened new opportunities for civic involvement. Guthrie appointed a Civic Commission to recommend improvements. The commission recommended new parks, playgrounds, roadways, sewers, and smoke control. Frederick Law Olmsted, Jr., was recruited to lead the new planning effort. Olmsted followed his father's pattern of using open space as a means of achieving a healthier city in all respects. Civic reformers strongly embraced this ethic (Bauman and Spratt 1996a, 154; Tarr 1996b, 170; Tarr 1989b, 241).

The commission interpreted its mandate as one of broad reform, including rail, water, and street systems, public lands and buildings, sewage, control over development, smoke abatement, and building codes. In sum, the initial desire of the commission was nothing less than a comprehensive city plan with a broad regulatory mandate overseen by impartial representatives. Olmsted, whose Pittsburgh efforts had included H. J. Heinz's home in Pittsburgh, was engaged to assist the commission's members (including Heinz) in their deliberations. Olmsted's report, published in 1911, was one of the more successful elements of the commission's works (Arnold, Freeman, and Olmsted 1909, 5). He recommended a new system of downtown roads and the conservation of Pittsburgh's steep slopes. He also recommended a major park at a 240-acre steep slope known as Nine Mile Run (Olmstead 1911, 119–120).

Olmsted's recommendations were followed insofar as they dealt with transportation. Otherwise, his public space recommendations were ignored. This was a remarkable failure. Olmsted's vision would have provided a variety of open space and aesthetic improvements to the region. It failed when Guthrie's administration was replaced once more by a Republican machine, run by a nephew of the original Magee. The new administration paid lip service to the reform agenda, but considered it Guthrie's plan and made no further use of the Pittsburgh Civic Commission. They particularly rejected the notion that planning and management were essential to the health and well-being of the city. Instead they seized upon the infrastructure recommendations and moved those forward (Baumann and Spratt 1996b, 166–167).

The Early Twentieth Century: Challenges of Environmental Reform

In 1911, the legislature passed a bill creating the Pittsburgh Department of City Planning. The department created an action plan for itself that included a review of the road system, an investigation of river improvements, and improvements to public buildings and lands including parks, adequate water supply, smoke abatement, and intergovernmental relations. But the Planning Department failed to create a comprehensive urban plan.

It also became increasingly apparent that the city was not willing to expend funds to keep open space in the city. Park acquisition became such a low priority that when the H. J. Heinz family offered to donate their East End home and landscaped grounds for a park, the gift was refused because of the maintenance costs. Frick Park had been accepted as a gift only a few years before, and the city was struggling to keep up its existing infrastructure, even though much of it had been a gift.

While Pittsburgh was unable to implement planning and land-use controls, other cities moved ahead. Frustrated by the inability to control industrial development, progressives found the situation increasingly intolerable. In 1918, banker Richard. B. Mellon urged civic and industrial leaders "to secure for the city the benefits of scientific planning such as is now being applied elsewhere in order to control its proper development, both industrially and socially." The group voted to form a citizen committee on planning (Pittsburgh Regional Planning Commission 1968, 4–6).

A subcommittee of this group quickly began to work toward overcoming the existing method of allocating parks and recreation in the city. They found that recreation was heavily tilted toward the wealthiest communities while working districts lacked access to open space. They also deplored the lack of available land for waterside parks. The subcommittee noted that in the whole Monongahela River valley only one site was adequate as a location for a waterfront park. All the others were foreclosed because of rail, road, or industrial activities at the water line (Citizens Committee on City Plan 1923, 11–15, 65).

In 1923, the committees recommended separate zoning for heavy and light industrial, commercial and low-density and high-density residential areas, and the limiting of building heights in certain zones. The act divided the city's real estate community and professional elites. Senator David Reed, testifying against the zoning ordinance, claimed the city's police power can only be applied for reasons of safety, health, and general welfare, and not for general civic beauty. "Beauty may be desirable, but liberty is more so," said Reed. "Neither the City Council nor the state legislature can exercise police powers for aesthetic reasons." The Planning Commission staff member responded that the ordinance carefully avoided aesthetic zoning, which is not authorized. Thus the zoning advocates were forced to argue that zoning was not an aesthetic issue but a control of nuisances. This severely limited their ability to keep industrial activity out of many parts of the city, especially open space (Pittsburgh Department of City Planning 1923; Pittsburgh Press 1923, 3–4, 20).

In this same period, Pittsburgh also attempted to implement air pollution control ordinances and create a citywide sewer authority. Neither effort succeeded. Smoke control ordinances were passed in the late nineteenth century in response to complaints from the Civic Club and other progressives, but the policy proved ineffective because there was no substitute for coal in transportation, industry, and home heating. Technology for cleaner combustion was limited, and cleaner fuels were not available. Smoke continued to choke Pittsburgh into the 1960s, when the attitudes changed and the availability of natural gas made cleanup possible (Tarr 1996a and 1996b).

The failure of progressive-era environmental reformers in Pittsburgh is significant and presaged many future problems. The coalition favoring environmental improvement did not seek any broader social improvement that might have taken labor support away from the Magee machine. Bureaucratic management, in the end, was not able to significantly stem the sea of waste and pollution that was enveloping Pittsburgh and the industrial heart of America.

The Pittsburgh Renaissance

The Second World War turned Pittsburgh's skyline black. Industrial production and lack of environmental controls made a bad situation intolerable. Streetlights were on twenty-four hours a day, and businessmen brought an extra white shirt to work every day. In 1948, approximately two dozen residents of Donora, Pennsylvania, were killed and hundreds hospitalized during a particularly bad temperature inversion over the town.

By 1945, Pittsburgh's civic and economic elite recognized that without improvements to Pittsburgh's poor environmental record—and a corresponding increase in the quality of life—the residents might abandon the city. Many of the goals of the reform movement, including widely touted improvements in air quality, were achieved in the 1940s and 1950s, in what became known as the Pittsburgh Renaissance. In lieu of the failed citizens' committee, Richard King Mellon and others in the business community brought all of the city's major industries together under the Allegheny Conference for Community Development (ACCD). The conference successfully partnered with Democratic Mayor David L. Lawrence on a reform agenda that included smoke control, housing rehabilitation, open space, and restoration of the central business district.

The result was a unique partnership, which for all its success, still suffered from the shortcomings of the earlier efforts. A strict smoke control ordinance, enacted in 1941, was aggressively implemented following the Second World War. Aiding its success was the transition from coal-fired steam engines to diesel-electric engines for rail transportation, natural gas for home heating, and increasingly efficient industrial combustion. Home heating represented the most significant change. Previously, Pittsburgh lacked any means to effectively regulate home-by-home use of coal; and there were few substitutes. With natural gas pipelines in the region, homes switched because it was economical to do so, and because low-sulfur home-heating coal was in short supply.

Railroads resisted the change to diesel-electric engines, but were persuaded to adopt it when pushed by R. K. Mellon, a board member of the Pennsylvania Railroad (Tarr 1988, 66–68).

The Renaissance also focused on downtown redevelopment, sewage control, and the creation of a park at Pittsburgh's point, where the Allegheny and the Monongahela rivers form the Ohio. More importantly, the region's major industries, civic organizations, and elected leadership collaborated in an arrangement that in large measure put the politics of the Progressive Era aside long enough to create a significant redevelopment of the central business district and a marked improvement in air and water quality (see figure 6.1). This reflected the widespread consensus that emerged in the years after World War II, that the dangerously polluted air was a significant barrier to future economic growth.

The private sector was essential to the Renaissance. The planning for the effort and the advocacy at the state and local level were carried out by the Allegheny Conference. The park at the Point and the high-rise Gateway Towers that replaced decades-old, dilapidated warehouses were financed both privately and publicly as a joint venture. The mayor's leadership helped to ensure public financing and more importantly, public support. The result was a significant improvement to the regional skyline. Unlike previous administrations, Mayor Lawrence was frequently willing to put politics and favoritism aside to pursue the restoration of the landscape.

Despite significant success, the partnerships created by the Renaissance still had limitations. They were elite organizations focused around major industries and lacking broad public involvement; that is not, however, to say they did not have support. Environmental improvement won support for the practical reason that the city might otherwise fail. With its sooty skies, increasing numbers of workers and businesses were considering other locales.

The Pittsburgh Renaissance was not completed in the 1950s, or even the 1960s. Point State Park, the former warehouse and rail yard site at the confluence of the rivers, was not fully completed until the early 1970s. Soot and air quality continued to be a problem—although nothing like the Second World War—into the 1970s. Sewer treatment has improved, but many systems in the outlying communities are still failing. While lasting partnerships between the public and the private sectors were created, a lasting commitment to environmental improvement was not.

Renaissance II

A second Renaissance began in the late 1970s, when Mayor Richard Caliguiri, business leader Henry J. Heinz II, and the Allegheny Conference sought to expand the downtown office and cultural districts and wipe up the remaining blight in the central business district. By and large successful, the second Renaissance resulted in a fourteen-block cultural district and a series of gleaming office towers. As historian Joel Tarr has pointed out, the second Renaissance was notable for its concentration of power in the public, rather than the private sector. The decline of U.S. Steel, Westinghouse, and other Pittsburgh corporations translated into less civic leadership and involvement, as a new generation of corporate leaders replaced the benevolent heads of such family-owned institutions as Mellon Bank. The city Planning Department, the Urban Redevelopment Authority, and other public institutions increasingly played the role of planner and financier, where Renaissance I had relied on the business community for such expertise.

One of the greatest challenges to Pittsburgh's efforts came from the concentrated period of economic decline, which accelerated during the years of Renaissance II. Manufacturing made up 50 percent of Pittsburgh's total employment in 1970. By 1990, it had fallen to 11 percent. More than 125,000 manufacturing jobs disappeared from western Pennsylvania in the space of a few years; outside of the former Soviet Union, few places on earth have undergone such profound economic change in time of peace. The result transformed the city. Prime riverfront lands became contaminated brownfields, often with rusted hulks of mills still present on-site. Skilled labor relocated; the city's population fell from 700,000 in 1950, to 360,000 in 1990. The economic working base of Pittsburgh was gone.

With regional decline came a loss of support for any environmental improvement that appeared to impose costs on the region. Economic opportunity and diversification became the goals of the 1980s and 1990s. Pittsburgh's air quality began to suffer, this time from auto exhaust and

Figure 6.1
Two photographs of Pittsburgh, before and after cleaning the air

coal-fired power plants in Ohio and West Virginia, and nagging problems such as sewer overflow, sprawl, and congestion began to plague the region. In 1993, the Commonwealth of Pennsylvania attempted to implement a centralized emissions inspection program to control smog from automobiles. Pittsburgh drivers rebelled, and the state legislature quickly repealed the program, even though it meant paying millions of dollars to the program's private contractors for breach of duty. A rumor that controls might be needed on backyard barbecues sparked a flood of calls to area radio stations protesting the audacity of the Environmental Protection Agency. Efforts to compel municipalities to separate storm water from sewer systems met with a dull indifference and noncompliance. In short, when environmental improvement was no longer a matter of changing corporate practices, it became a much lower priority.

It should come as no surprise that changing individual behavior would be a problem in Pittsburgh. Since the late nineteenth century there has been a disconnect between public-private partnerships that seek to reshape the central business district, and the larger Pittsburgh region.

Figure 6.1
(cont.)

Because of Pittsburgh's pattern of elite political and economic actors, there has been a weakness in the civic infrastructure. There are very strong neighborhoods and local organizations; but there are only a few citizens' groups capable of acting communitywide. There is no definitive environmental voice in the region; rather, there are several. In short, this disconnect between high-profile, often successful partnerships and day-to-day Pittsburgh is a barrier that the effort to further environmental improvement will have to overcome.

A New Vision for Pittsburgh

In the early 1990s, the leaders of the city and region once again collaborated to try to revitalize the economy.[1] The Allegheny Conference, noting the lack of consensus among key stakeholders concerning the economic future, appointed a task force in 1993, to undertake a comprehensive self-assessment of the region. Led by Robert Mehrabian, president of Carnegie Mellon University, the six-month study involved careful benchmarking of other U.S. cities of a similar size and character (Mehrabian et al. 1993). The results were sobering. They revealed that, despite its many remaining assets, Pittsburgh had experienced the slowest employment growth of any comparable city in the nation, and faced a bleak future unless the community mobilized immediately to build a new economic base.

The Allegheny Conference responded to this challenge by putting in place a phased economic revitalization process (it came to be known as the RERI, for Regional Economic Revitalization Initiative) that had several stages. The first involved an extensive process of consensus building, focusing on the need for a shared vision and collective action in response to the region's newly defined problem of economic stagnation. A leadership group created by the conference held numerous outreach sessions, engaging more than 5,000 citizens in discussion groups. This culminated in a special live television broadcast to publicize the initiative and build commitment to action (Allegheny Conference on Community Development 1994).

By late 1994, the RERI had led to the formation of an action group called the Working Together Consortium. Created under the leadership of Allegheny Conference Chairman Thomas O'Brien, and then carried forward by his successor, Richard P. Simmons, the consortium set forth an ambitious agenda of collaborative activities aimed at rebuilding the regional economy. The most widely publicized goal was the creation of 100,000 new jobs by the end of the century, but the consortium also set forth an ambitious agenda of steps needed to achieve this. Among the most concrete of the latter was the formation of a special $40 million Strategic Investment fund to underwrite revitalization activities. The fund, which began operations in 1996, took the form of a limited partnership with corporations, foundations, and even some individuals, investing.

The consortium also created approximately eleven special committees made up of volunteer leaders from every sector to implement the action recommendations agreed upon by the RERI. The committees agreed to report on progress at an annual Competitiveness Summit. Finally—and perhaps the most distinctive feature of the process—all of the activities were to be monitored and evaluated annually by one of the region's principal accounting firms, Deloitte & Touche, LLP, as a community service (Allegheny Conference on Community Development 1996).

The new initiatives became so numerous that many participants had trouble keeping track of them. Taken together, the RERI process, the formation of the Working Together Consortium and the Strategic Investment Fund and the annual summits, surely represent one of the most ambitious and comprehensive community planning and economic action programs ever undertaken in an American city. Certainly it is distinctive in its intensity and in the degree to which it involved a very wide participation of leaders from every sector, public and private, in the community.

The environment was not at the forefront of this effort, nor was what the readers of this book would consider "sustainable development." To be sure, the program did include a commitment to reclaim the riverfronts—riverfronts marred by thoughtless industrial development and dominated by shabby railroad tracks. But interest in the rivers, and parks along them, was driven mainly by aesthetic considerations and a realization that the professionals working in new, high-technology industries were becoming accustomed to a far "greener" urban context than Pittsburgh could offer. Central concepts that readers of this book would identify as key to an understanding of sustainable development were simply not part of the revitalization process. Key leaders at the conference were not interested in these things as they set forth the development vision of the region. The objective was economic growth, as quickly as possible, and environmental trade-offs were acceptable to the business leaders who were in charge of the process.

Adding to the Vision: The Environmental City Network

As the multiple initiatives of the economic revitalization process were gathering steam, a coalition of business, foundation, and academic inter-ests met with the mayor to discuss the environment. This coalition evolved primarily from a partnership of key leaders of the so-called green business community—who had formed an environmental business net-work to look after their interests—and local organizations, including the Heinz Endowments, a Pittsburgh-based private foundation and its grant-ees, which sought to build a sustainable community. The mayor's goal was to promote economic development. In his travels to other cities, he had observed their plans to utilize the "environment" as a mechanism of stimulating economic growth, so he convened a group of liked-minded parties to explore the possibilities. His desires complemented those of several local foundation leaders, as well a number of academic specialists, all of whom had become increasingly alarmed that key private-sector leaders—primarily in the Allegheny Conference—apparently had little interest in issues associated with sustainablility as it is understood in this book.

Convinced that Pittsburgh could take advantage of the growth of environmental commerce, Mayor Murphy in 1995, commissioned a task force to see whether the city could take a leadership role in this field. The five-month study revealed the existence of a rich network of small envi-ronmental companies, most of them the product of many years of cleanup and remediation, as well as a range of environmental research and teach-ing activities at the region's major universities.

The task force produced a strategic report and a blueprint for what has become known as the Environmental City Network (ECN)—a part-nership of public-sector, private-sector, and foundation initiatives aiming to unite the region's environmental efforts. The network's mission is to "foster sustainable development, environmental expertise and economic growth in southwestern Pennsylvania in a manner that enhances the region's quality of life and business opportunities for future generations."

Building on the city's environmental expertise, the network has four major divisions in its structure: Environmental Business, the Public Sector, Community-Based Organizations and the University Community. Some of the network's activities will include:

- Land-use planning
- Riverfront development
- Environmental education and research
- Business development and sustainable business activities
- Community stewardship
- Marketing

Each of the clusters of activity that form part of the Environmental City Network has an existing base of support, but the parallel efforts of each cluster have never been combined to create a competitive advantage. The programs discussed below are examples of multistakeholder initiatives being developed in the region, often from the grassroots level.

Environmental Knowledge

To foster the growth of research and technology commercialization in a variety of fields related to the environment, the ECN sponsored the formation of an alliance named The Three Rivers University for the Environment. This is an academic partnership made up of the region's major universities—Carnegie Mellon, the University of Pittsburgh, Duquesne University, and Chatham College—to integrate academic programming and environmental research, and offer a unique forum within which to share and strengthen programs in environmental studies. The alliance is an effort to educate future generations of experts, and create opportunities for scholars from around the world to carry forward the responsibility of environmental stewardship.

Environmental Renewal

Brownfields Pennsylvania is home to more than 10,000 contaminated industrial sites, many in the Pittsburgh area. Addressing these brownfields and the obstacles they pose to economic development is one of the city's greatest challenges. For many former industrial sites, the true impediment to reuse is not cleanup itself, but uncertainty about the scope of such an undertaking. These sites could become attractive for a variety of new uses if the total environmental development cost for assessment and remediation were already known. Resolution of questions about environmental concerns allows the sites to be evaluated more promptly and favorably by potential purchasers. Moreover, the sites also offer genuine advantages to prospective new purchasers, such as existing transportation infrastructure, established utility service, and nearby workforce and consumers.

Nine-Mile Run A partnership between the ECN and the city of Pittsburgh has been formed to support the design of public greenspace next to the largest undeveloped brownfield site in the city of Pittsburgh. The city has also begun planning for a new housing development adjacent to the proposed greenway. Nine-Mile Run, buried under 150 feet of steel slag and stretching 240 acres to the Monongahela River, is an area of real estate that has been dormant for nearly a century. The project aims to turn the massive slag pile into a thriving urban community by showcasing innovative solutions and addressing the full range of development challenges for urban brownfield sites.

Phoenix Land Recycling Company The Phoenix Land Recycling Company is a unique nonprofit corporation that facilitates the assessment and remediation of brownfield sites for economic reuse. The organization partners with local economic and industrial development agencies, commercial realtors and developers to identify locations with a high potential for economic reuse. Phoenix conducts comprehensive assessments and develops remediation plans for these sites, and with the requirements and costs quantified, it assists in marketing the sites to prospective purchasers. Phoenix is also working with the Pennsylvania Department of Environmental Protection, which is committed to establishing a model land-recycling program throughout the state. Its first successful brownfield recovery project was nearing completion at this writing.

Pittsburgh RISES The Pittsburgh RISES Project is a collaborative project between the University of Pittsburgh and Carnegie Mellon University designed to attract companies to old industrial sites by creating and maintaining a centralized, graphically based inventory of available industrial property. It provides the means to conduct custom physical, economic, and environmental analyses, so informed decisions and comparisons can be made by prospective developers, planning organizations, community groups, and policymakers.

South Side Local Development Corporation In an attempt to demonstrate another way to meet the challenge of brownfield development, the South Side Local Development Corporation (SSLDC), a nonprofit developer in Pittsburgh, is building environmentally advanced housing on one of the more visible of these sites—the barren wasteland that was once home to U.S. Steel's South Side steel mill. The plan, dubbed the New Birmingham Project, will offer moderate and middle-income homebuyers the opportunity to purchase a home in the city's first cost-effective environmental housing development.

Riverfront Development
With the collapse of the steel industry, long stretches of riverfront land were abandoned—many of them seriously contaminated. The City of Pittsburgh Planning Department, in partnership with a broad cross-section of other organizations, is developing a comprehensive riverfront development plan. The objective is to articulate a vision for all riverfronts in the city, detail the projects currently under way, and identify the resources it will take to make the various development plans a reality. The plan will reflect the best practices of riverfront development based on the experiences of cities throughout the country. The Three Rivers Heritage and Steel Heritage Trails are expected to provide significant contributions to quality of life and economy. In addition, the Wildlife Habitat Council has launched the Three Rivers Heritage Partnership, in order to restore the riparian zones and recreate essential habitat

Green Neighborhoods
The Green Neighborhood Initiative is a program that targets low- to moderate-income neighborhoods for significant energy, water, and resource savings in order to increase household income and spur neighborhood redevelopment. Managed by the local nonprofit Conservation

Consultants Inc. (CCI), it is currently operating in several Pittsburgh neighborhoods and has expanded its focus to include lead risk reduction and open-space programs for participating neighborhoods. It has already produced tangible results in two communities—Carrick and Tarentum—where the program has helped to reduce pollution, improve business profits, and increase the value of local housing.

In 1996, the town of Carrick became the first "green" neighborhood in Western Pennsylvania, a move prompted by a coalition of public, private, and community organizations striving to revitalize the area by cutting residential and business energy costs. Their hope is that money saved through energy conservation measures, including weatherization and recycling, will be reinvested in the community. Over 50 percent of Carrick businesses and residents are participating in the Green Neighborhood Initiative. Those involved have already exceeded the program's initial expectations by saving an average of 30 percent in annual electric, gas, and water bills since the program began.

The town of Tarentum recently decided to reassume control of its electric utility after leasing its system for twenty years. To facilitate the transition, the town issued an urgent plea for assistance in reducing energy costs. CCI stepped in to provide weatherization assistance to low-income households, and audit and consulting services for several municipal buildings in the town. It also convinced local banks to offer energy efficiency loans to the town's middle-income residents and businesses. Tarentum homes, many of which have utility bills exceeding monthly mortgage payments because of the age of the buildings, are now being rehabilitated to maximize energy and resource efficiency by combining utility weatherization programs with low-interest financing from area banks. The homes that are participating in the program have saved an annual average of 35 percent in gas bills—the highest in the state of Pennsylvania.

Green Buildings

Last year, the Western Pennsylvania Conservancy created the first "green" building in downtown Pittsburgh. The organization was looking for an opportunity to create a sustainable, environmentally sound location for its new headquarters, and chose to retrofit a historic downtown building to showcase energy efficiency and green redesign. The conservancy is also working to promote public awareness of cost-effective, environmental building principles through an exhibition in the new ground floor resource center. A number of local organizations were involved in providing technical support for the project, including Conservation Consultants, Carnegie Mellon University, and the Green Building Alliance.

Land Use
Reestablishing population density in the urbanized areas is currently a priority of city government. Committed to attracting the middle class back into the city, the mayor and his housing deputies are focusing on building housing communities within the urban core and rehabilitating the existing housing stock. Successful models include Crawford Square, South Side, Washington's Landing, and the North Side—some of which are on former industrial sites.

Environmental Enterprise
Southwestern Pennsylvania has a surprisingly high concentration of environmental firms. Over 800 Pittsburgh companies are currently engaged in environmental business regionally, nationally, and throughout the world, and environmental firms have been identified as one of the key clusters of economic growth in the region.

The Environmental Business Network   Under the Pittsburgh High Technology Council, the Environmental Business Network (EBN) is also building an alliance to promote the region's environmental technologies and manufacturing processes in domestic and international markets. The EBN is working in cooperation with the World Trade Center of Pittsburgh, the Pennsylvania Department of Community and Economic Development, and the regional office of the U.S. Department of Commerce to create major export opportunities for environmental service companies.

Sustainable Industry   A key goal of the Environmental City Network is to foster an industry that reduces the use of toxins and looks toward energy efficiency, waste minimization, and pollution prevention. Out-dated organizational designs and the challenges to improving production facilities often cost companies unnecessary time and money. Although most of these manufacturers are aware of and even complying with environmental regulations, many do not know about the economic benefits of reducing pollution at the source rather than mitigating its effects. In addition, even fewer companies are aware of the significant cost savings associated with adopting energy efficient practices.

Duquesne University To respond to this challenge, Duquesne University's Institute for Economic Transformation is working to create regional and statewide industrial networks committed to pollution prevention and efficient use of resources to improve environmental performance. Duquesne is collaborating on this project with the Business for Social Responsibility, a San Francisco based not-for-profit organization dedicated to improving economic and social performance in all industries.

SPIRC To assist regional manufacturers, the Southwestern Pennsylvania Industrial Resource Center (SPIRC), an affiliated organization of the Pittsburgh Technology Council, has developed a methodology to assess energy, waste, and productivity in local manufacturing facilities.

SPIRC is addressing a significant concern of regional manufacturers: improving environmental performance in ways that increase productivity and improve the overall economic performance of the companies. Ultimately, the program strives to strengthen the region's capacity for growth while promoting environmentally aware business practices.

Alternative Energy A number of green technologies and alternative energy sources are being developed in the region that will shape future energy choices. These include the development of a consortium to develop hybrid-electric engines that serve as a substitute for internal combustion engines in niche markets. In addition, the Pittsburgh region has actively worked for the development of fuel cells in both public and private research facilities.

Partnerships have been essential to both efforts. The federal government has supported fuel-cell research in Pittsburgh since the 1970s. Area utilities, including Duquesne Light, Equitable Gas, and Consolidated Natural Gas, have made strategic investments in fuel cells, and a demonstration cell now provides power at the Pittsburgh airport.

Other Sustainable Initiatives in the Pittsburgh Region

Sustainable Development   The Pittsburgh chapter of the American Institute of Architects' Committee on the Environment has developed a plan for creating a sustainable future for Western Pennsylvania. The project identifies the region's strengths and focuses on the challenges of creating a sustainable community.

The project is bringing together stakeholders—including municipalities, citizens, and community and environmental groups—to foster greater cooperation and encourage better planning, resulting in a greater quality of life. The objective is to design more sustainable plans for land use and growth patterns. One of the first projects includes advocacy to adopt regulations that will result in higher standards for private development in one of the region's fastest growing corridors, the Parkway West.

Community and Economic Development   Pittsburgh and Southwestern Pennsylvania have a strong community economic development movement. Community development corporations exist in more than fifty communities in distressed areas in the city and former mill towns. Their commitment to revitalizing their declining areas has resulted in the creation of thousands of square feet of commercial and residential properties, the development of new businesses, and the gradual stabilization of communities that might otherwise have been lost. Revitalization activities mounted by these small organizations are often the only efforts under way to turn around distressed communities.

The community-based component of the Environmental City Network will provide a network from which public opinion can be harnessed in order to provide for better-informed public input into community processes. It will work to educate the public and increase participation in activities that support, protect, and restore southwestern Pennsylvania's natural and built environments.

Regional Marketing and Economic Development   Building on the efforts of the Allegheny Conference, the Pittsburgh Regional Alliance (PRA) was created to coordinate local economic development efforts among six agencies: the Greater Pittsburgh Chamber of Commerce, Penn's Southwest, the Regional Industrial Development Corporation, the World Trade Center, the Pittsburgh Technology Council, and the southwestern Pennsylvania Industrial Resource Center. PRA's mission is to nurture the startup and growth of new enterprises, help existing business grow and expand, and retain them in the region, and market the region and attract business from outside.

The Environmental City Network is currently working to develop an environmental marketing plan for the region. In coordination with the efforts of the PRA, the ECN is working to spread the message of Pittsburgh's environmental accomplishments. It is developing a national marketing campaign to change the impression and image nonresidents have about the quality of life in southwestern Pennsylvania. It is also working to create public education materials (including a Web site) to enhance environmental literacy and stewardship in the area.

Conclusions

The Pittsburgh case provides a useful laboratory for the study of environmental transitions at the local and regional levels in the United States. When examined with the conceptual framework of this book, it provides a basis for optimism about the prospects for local initiative. However, it also provides a sobering acknowledgment of the constraints that can be expected to limit the move to a more sustainable future without strong policy frameworks operating concurrently at the national and even international levels.

The leaders of Pittsburgh and Allegheny County dramatically reduced local air pollution in the years immediately following World War II, and they did so in a decisive and effective manner. The cleanup was part of a collaboration between New Deal politicians and conservative businessmen. It was driven partly by a growing concern about the impact of smoke and ash on health and quality of life in the region, but mainly by a conviction that the local economy (in deep trouble) could not improve without a range of actions to revitalize the city (Tarr 1996b, 228). Pittsburgh had a midwestern model to follow—the widely touted St. Louis cleanup—and was lucky to have strong and politically secure leadership in place. The costs of the cleanup were real, and were borne to a significant extent by the poor, who faced fuel shortages and eventually had to pay the costs of conversion to natural gas (Tarr 1996b, 249). These actions were firmly within the regulatory, or epoch-one model of environmental action, as set forth in chapter 1.

From the mid-1950s to the early 1990s, Pittsburgh remained frozen in time as far as local initiatives affecting the environment were concerned—it was still, by and large, an epoch-one city. Environmental issues had no place in the political dialogue or in the concerns of regional leaders, as the city's economy (despite an office tower building boom) rapidly declined with the demise of the steel industry and the exodus of the middle class to the suburbs. The economic problems of the region were so severe that they overrode all other concerns, including those related to the environment. Pittsburgh's cleanup predated federal mandates, and thus it did not face the difficulties that more recalcitrant communities dealt with in the 1970s and 1980s.

Pittsburgh made a new commitment to environmental action in the mid-1990s with the announcement of the Environmental City Network by the mayor, county authorities, and a coalition of foundation and nonprofit leaders. The first and most important reason for the network was the belief that the promotion of environmental business would help revitalize the economy. City leaders, most notably the mayor, envisioned a rapid expansion of environmental companies that would employ citizens and generate international exports. In this respect, the motive was the same as it had been in the cleanup of the smoky city a half-century earlier.

In addition, benchmarking by Pittsburgh leaders—searching for keys to success in building a new economy—had revealed that the "look" of those cities most successful in attracting young, entrepreneurial, high-tech workers was often very "green." Parks, bike paths, greenways, clean air, trees, access to rivers—all the things Pittsburgh leadership had neglected—were attributes of cities successfully building the small, technology-based companies of the future. Green seemed to be the color of success in a number of other cities, and it was a label Pittsburgh hoped it could adopt at little or no cost.

The second motivation was the commitment of a handful of leaders in the university and foundation community to the vision of a city that was both environmentally friendly and had the potential to become a dynamic place of origin for sustainable economic activities of the future. These leaders hoped that such activities would go beyond measurement and remediation of pollution to include such things as design for pollution prevention and disassembly, and energy efficient transportation. For a handful of these leaders, the picture of a sustainable Pittsburgh, one building an epoch-three community, was a compelling motivation and an authentic one, as far as sustainability is concerned.

Unfortunately the ECN, and the partnership that carried it forward, was far different from the partnership that cleaned up the smoky city after World War II. Pittsburgh city leaders hoped that the green symbolism and the endorsement of environmental business would suffice; the largest firms and the most powerful business leaders (for the most part) have paid little attention. The main impetus and funding for the network came from nonprofit and foundation advocates, academics, and companies making up only a subcluster within the regional economy. A confirmation of marginality came when the network was not mentioned as part of the grand design for the future of the city announced with great fanfare by the mayor, regional authorities, and the business community a few short months after the ECN itself was launched.

Despite the tentativeness of these first efforts, the variety of initiatives begun in the mid-1990s by those committed to a vision of sustainable Pittsburgh was impressive. They ranged from efforts to improve housing energy efficiency and to promote "green" building practices, and extended to the introduction of several ambitious collaboratives to advocate for everything from research to environmental business networking. Foundation funds were critically important in getting the initiative started, and they will be needed for several years as the momentum develops. Early signs provide the basis for optimism concerning the emergence of a critically important ethos of sustainable development—an ethos that will be needed for the movement to become self-sustaining.

Although its historic accomplishments were bold and effective, by modern standards, Pittsburgh is not yet an "environmental city," nor is it a place where "environmental sustainability" is well understood or valued. A great deal of work remains, but a transition is clearly under way.

It is possible to suggest, from the Pittsburgh experience, a number of conditions that may support the emergence of a transformative effort such as the one that occurred after the Second World War (the Renaissance) and the one that is now under way:

• Decline of basic industries
• Failure of traditional economic development strategies
• Demand for an improved quality of life (including less pollution, greater recreation, and higher expectations)
• Emergence of a cluster of civic leaders and institutions seeking change
• Support from enlightened foundations

In many respects, Pittsburgh is a remarkable place. It is a city of strong neighborhood and regional loyalties, and it has a distinguished history of cooperative action to benefit the community. Among American cities, Pittsburgh will always hold a place in history as one that rose from the worst environmental conditions by the bootstraps. Pittsburgh leaders acted in a decisive manner, with New Deal Democrats cooperating with the most powerful conservative business leaders, to clean up the city. We have no doubt that the ECN will, in time, bear significant fruit. In a decade, Pittsburgh will surely have a more robust group of environmental companies, the research products of its universities will focus more on matters important to sustainability, and it will have more parks, trees, and riverfront amenities. With luck it will be better positioned to take advantage of new economic activities that will contribute to a sustainable economy in the future. And the ECN will build a foundation for even more decisive action favoring sustainability.

Note

1. The collapse triggered by the decline of the steel industry was, by this time, nearly complete. Growth in other sectors of the economy had failed to compensate for the losses in manufacturing. Between 1970 and 1990, average annual

wages actually declined by more than $500 per worker. The public sector, starved for tax revenues, postponed badly needed infrastructure investments. And, perhaps most important, the city and region seemed to be adrift, with no real solutions in sight.

References

Allegheny Conference on Community Development. 1996. The Greater Pittsburgh Region: Working Together to Compete Globally: Progress Report prepared for the Working Together Consortium, First Annual Competitiveness Summit, Wednesday, April 10, 1996 (Spring).

Allegheny Conference on Community Development. 1994. The Greater Pittsburgh Region: Working Together to Compete Globally (November).

Arnold, Bion J., John R. Freeman, and Frederick Law Olmsted. 1909. *City Planning for Pittsburgh.* Pittsburgh: The Pittsburgh Civic Commission.

Bauman, John F., and Margaret Spratt. 1996a. "Civic Leaders and Environmental Reform." In *Pittsburgh Surveyed,* ed. Greenwald and Anderson.

Bauman, John F., and Margaret Spratt. 1996b. "The Pittsburgh Survey and Urban Planning." In *Pittsburgh Surveyed,* ed. Greenwald and Anderson.

Byington, Margaret. 1996. *Homestead: The Households of a Mill Town.* Pittsburgh: The University of Pittsburgh Press (reprint of 1910 work).

Citizens Committee on City Plan. 1923. *Parks Report.* Pittsburgh: Citizens Committee on City Plan.

Greenwald, Maurine. 1996. "Visualizing Pittsburgh in the 1900s." In *Pittsburgh Surveyed,* ed. Greenwald and Anderson.

Mehrabian, Robert et al. 1993. Toward a Shared Economic Vision for Pittsburgh and Southwest Pennsylvania. Pittsburgh: Allegheny Conference on Community Development, November.

Olmsted, Frederick Law, Jr. 1911. *Main Thoroughfares and the Down Town District.* Pittsburgh: Pittsburgh Civic Commission Report Number 8. Harrisburg, PA: Mount Pleasant Press.

Pittsburgh Regional Planning Commission. 1968. *Prelude to the Future.* Pittsburgh: Pittsburgh Regional Planning Commission.

Pittsburgh Department of City Planning. 1923. *Arguments against the Proposed Zoning Ordinance.* Pittsburgh: Department of City Planning.

Pittsburgh Press. 1923. "Zoning Bill to Govern Building Operations Here Sent to Council," *The Pittsburgh Press* (January).

Reiser, Catherine Elizabeth. 1951. *Pittsburgh's Commercial Development.* Harrisburg, PA: Pennsylvania Historical and Museum Commission.

Steffens, Lincoln. 1906. *Shame of the Cities.* New York: McClure's.

Tarr, Joel A. 1988. "Four Decades of Public-Private Partnerships in Pittsburgh." In *Public-Private Partnerships in American Cities*, ed. R. Scott Fosler and Renee A. Berger. Lexington, MA.: Lexington Books.

Tarr, Joel A. 1989b. "Infrastructure and City Building." In *City at the Point*, ed. Hays.

Tarr, Joel A. 1989a. "Infrastructure and City Building." In *City at the Point*, ed. Samuel P. Hays. Pittsburgh: University of Pittsburgh Press.

Tarr, Joel A. 1996b. *Search for the Ultimate Sink: Urban Pollution in Historical Perspective*. Akron, OH: University of Akron Press.

Tarr, Joel A. 1996a. "The Pittsburgh Survey as an Environmental Statement." In *Pittsburgh Surveyed*, ed. Maurine Greenwald and Margo Anderson. Pittsburgh: University of Pittsburgh Press.

7

ISTEA and the New Era in Transportation Policy: Sustainable Communities from a Federal Initiative

Thomas A. Horan, Hank Dittmar, and Daniel R. Jordan

In 1960, a Professor at Syracuse University named Daniel Patrick Moynihan published an article entitled "New Roads and Urban Chaos" (Moynihan, 1960). In it, he questioned whether the Interstate and Defense Highway program, a federal plan initiated in 1956 to construct a 44,000-mile nationwide system of highways, was destined to "bring about changes for the worse in the efficiency of our transportation system and the character of our cities." Particularly troubling to Moynihan was that an endeavor of such magnitude was proceeding with so little thought of its consequences for communities across America. "Highways determine land-use . . . [and] the spreading pollution of land follows the roads," argued Moynihan, "[so] it stands to reason that engineers should be required to conform their highway plans to metropolitan land-use plans designed in the context of more general economic and social objectives." But no such planning was occurring, thus "the [Interstate] program is doing about what was to be expected: throwing a Chinese wall across Wilmington, driving educational institutions out of downtown Louisville, plowing through the center of Reno." Moynihan called for more oversight of the Interstate Highway program, comprehensive metropolitan planning, and a reappraisal of America's need for large-scale highway construction. These actions, he hoped, might "impart some sanity and public purpose to this vast enterprise."[1]

Thirty-one years later, Daniel Patrick Moynihan, now the prominent Democratic senator from New York, was witnessing President George Bush signing the Intermodal Surface Transportation Efficiency Act of 1991 (known as ISTEA, pronounced "Ice-Tea"). A chief author of ISTEA, Moynihan declared the landmark, six-year, $151-billion surface transportation bill "the first legislation of the post-Interstate era. We have poured enough concrete. Our primary objective now must be to improve . . . the system we have" (Moynihan 1991).

Politicians in both parties, the popular press, and academics similarly trumpeted ISTEA's significance. And advocacy groups promoting the environment and community interests heralded ISTEA's unprecedented effort to balance multiple transportation policy objectives, such as mobility, environmental quality, social equity, and the preservation and enhancement of areas of cultural and aesthetic value.

ISTEA's significance, in short, went far beyond providing more money for roads and bridges. The legislation represented the most serious federal effort yet to articulate and institutionalize a "holistic" vision of transportation policy, a vision consistent with principles labeled in this volume as "sustainable communities."[2] Perhaps the central premise of ISTEA's holistic approach was that transportation touches the lives of everyone, plays a critical role in determining the "livability" of communities, and has enormous consequences for the natural environment. This makes post-ISTEA transportation policy an excellent case for studying the principles described in chapters 1 and 2 because it provides a revealing example of how a policy area, having philosophically transitioned into the sustainability epoch in many important respects, is faring in the implementation of these new principles.

Is ISTEA's holistic approach in fact being implemented? Has this new approach had any discernible effects on the transportation system? What lessons does the post-ISTEA experience provide as other policy areas enter the sustainable communities epoch? To what extent has subsequent legislation—principally the Transportation Equity Act for the 21st Century("TEA-21")—carried forth this mandate for a more integrated approach to transportation planning and financing?

In seeking answers to these questions, we focus first on the interplay of transportation and environmental policy over the last thirty-five years. This historical perspective is useful, for it demonstrates the considerable extent to which the environmental movement influenced federal transportation policies prior to ISTEA. Indeed, at least at the level of federal statute, some of the sustainability thinking considered new in ISTEA and characteristic of the third environmental epoch in fact existed in federal transportation policy as early as 1962. ISTEA's key contribution, however, was to refine and consolidate these ideas and provide ways to actually implement them.

The remainder of the chapter analyzes the implementation of ISTEA in three areas: governance, planning, and technology. Two themes stand out as these three policy areas are explored. First, there remains a good deal of ambiguity surrounding precisely what the "sustainable communities epoch" (as described by Mazmanian and Kraft in chapter 1) implies for transportation. The relatively simple goals that once justified transportation investments (i.e., moving traffic as quickly as possible) have been replaced by multiple goals that are subtle, complex, and often vague (i.e., investments should improve a community's overall "quality of life"). This complicates the task of understanding whether particular transportation investments comport with sustainability principles.

Second, despite ISTEA establishing the national framework within which transportation decision are made, the law provided no guarantees that state and locally implemented transportation projects will reflect its holistic approach. ISTEA, while strengthening the hand of sustainability advocates, in no way assured that these advocates will prevail during implementation (Gifford, Horan, and White 1994). As the chapter will show, the sustainable communities epoch has had far greater impact in some areas of transportation than it has in others. Nonetheless, the passage of key elements in ISTEA and their retention in TEA-21 places surface transportation in an important role to contribute to the achievement of the third epoch in environmental policy.

Policy through the Rear-View Mirror: Three Decades of Emerging Environmental Concerns

For most of this century, federal surface transportation policy focused on accommodating the demand for travel and the needs of automobiles by promoting the construction of roads and other transportation facilities (Giuliano and Wachs 1992; Cervero 1989). This policy reached its zenith with the Federal-Aid Highway Act of 1956, which provided a steady source of funds to construct the Interstate Highway System. The act created the Highway Trust Fund to collect revenues from highway user fees (i.e., gasoline and diesel taxes, tires, truck sales, and highway use), and distribute these revenues for Interstate Highway construction projects. To encourage highway building, the federal government covered 90 percent of the construction costs for Interstates, while states paid the remaining 10 percent. The large amount of money dedicated to highways narrowly focused federal and state activities on building the Interstate System.[3]

Interstate construction proceeded rapidly but not without controversy. By the late 1960s, over two-thirds of the system was completed (Leavitt 1970). The pace of construction, and the environmental and community disruption often accompanying it, prompted citizen groups and local governments opposed to highways to mobilize throughout the country during the 1960s.[4] It was also during this period that federal policy began slowly to change. While the dominant policy goal remained Interstate construction and the Highway Trust Fund assured a continuing stream of funds for this purpose, federal attention began focusing more on the broader social and environmental consequences these projects were having. This new focus manifested in transportation laws, as well as in environmental laws aimed partially at transportation, passed between 1962 and 1990 (see table 7.1).

Among transportation legislation during this period, most significant was the Federal-Aid Highway Act of 1962. The 1962 act marked a philosophical watershed: it explicitly acknowledged the link between transportation and broader community and environmental values, and called for integrated policy approaches to address these broad issues. In doing so, it foreshadowed elements of ISTEA's holistic framework and was a harbinger of the sustainable communities epoch. Specifically, it called for the creation of a regional transportation planning process in which planners had to analyze a series of ten "elements" in devising their transportation plans. These elements included the effects of transportation projects on land use, the environment, and on other "social and community-value factors," such as open space and aesthetics. This philosophical change, however, made little difference in practice, as federal transportation projects remained focused on highway construction and traffic efficiency (Rose 1990; Morehouse 1971; Leavitt 1970). Simply put, the 1962 act provided neither the financial resources nor the overall implementation structure that are necessary conditions for effective implementation (Mazmanian and Sabatier 1989).

Table 7.1
Select federal legislation after 1956, addressing the social and environmental impacts of transportation

Legislation	Important provisions
Transportation legislation	
1962 Federal-Aid Highway Act	Mandated "3C" transportation planning process; created ten "planning elements" for consideration in the transportation planning process (i.e., land use, social and community values, etc.)
1966 Department of Transportation Act	First federal statutory language aimed at reducing the effects of highway construction projects on the natural environment: prohibited "the use of land for a transportation project from a park, recreation area, wildlife and waterfowl refuge, or historic site unless there is no feasible and prudent alternative and the project is planned in a manner as to minimize harm to the area."
1968 Federal-Aid Highway Act	Required that public hearings be held on proposed highway projects; created the highway beautification program and provided relocation assistance to those displaced by highway projects.
1973 Federal-Aid Highway Act	Allowed unprecedented flexibility in the use of Highway Trust Fund for mass transit
Environmental legislation affecting transportation	
1969 National Environmental Policy Act	Created "Environmental Impact Statement" process
1970 Clean Air Act	Created emissions standards and mandated the establishment of national ambient air quality standards (NAAQS)
1977 Clean Air Act	Instituted "conformity" regulations and sanctions for noncompliance with air quality standards; relaxed some compliance deadlines; required transportation control measures
1990 Clean Air Act	Strengthened "conformity" regulations and tightened emissions standards

Perhaps more consequential in practice for transportation during this period were numerous environmental statutes. Two in particular had profound implications for transportation policy—the National Environmental Policy Act of 1969 (NEPA), and the 1970 Clean Air Act (CAA). NEPA focused on process and information, mandating that all federal agency projects "significantly affecting the environment" include an Environmental Impact Statement (EIS).[5] Perhaps no other federal activity has felt NEPA's influence more than transportation: between 1979 and 1992, the U.S. Department of Transportation (DOT) filed more EISs than any other federal agency—from as many as 277 in 1979, to as few as 80 in 1989. And in 1991, DOT was named as the defendant in NEPA claims more often than any other federal agency (Kraft 1996). The Clean Air Act of 1970, conversely, adopted a prescriptive, "command-and-control" approach to regulating transportation. It required, for example, automobiles to reduce certain emissions by 90 percent by either 1975 or 1976.[6] The CAA amended in 1977, and again in 1990, with the 1990 amendments significantly expanding provisions related to "conformity" determinations in areas failing to meet air quality standards.[7]

The transportation and environmental legislation passed between 1960 and 1990 places the significance of ISTEA into perspective. Measured against laws instituted during the previous decades, ISTEA was not a revolution in ideas. The notion that transportation should serve broad social, economic, and environmental goals appeared in federal law in 1962. What was revolutionary, however, was ISTEA's restructuring of U.S. transportation policy in ways that gave new "teeth" (i.e., implementation tools) to the old directive that transportation investments be environmentally sensitive and cognizant of community values. These new teeth appeared in several areas. We first explore provisions that strengthened the role of regional agencies known as "metropolitan planning organizations."

New Drivers on the Road: The Rise of MPOs in Transportation Governance

The extension of local control over policy decisions is a distinguishing characteristic of the sustainable communities epoch. ISTEA did this in transportation by expanding the role of metropolitan planning organizations (MPOs) in the transportation policy-making process. In all urbanized areas with populations exceeding 50,000, ISTEA made MPOs responsible for devising short- and long-term transportation investment plans, for selecting which transportation projects to fund within their jurisdiction, and for ensuring that transportation investments are consistent with air quality laws and regional growth management strategies (see table 7.2).

Metropolitan planning organizations (MPOs) are diverse, often organizationally complex entities. They differ in size, technical capabilities, and degree of political independence (U.S. Advisory Commission on Intergovernmental Relations 1995, hereafter ACIR 1995). While all MPOs derive their authority under federal urban transportation planning mandates, some possess additional state-conferred powers to address regional growth management and economic issues. Some have large staffs and considerable in-house expertise to conduct transportation planning, while others have few resources and depend largely on state DOTs to carry out their responsibilities. The roughly 339 MPOs designated under ISTEA were generally alike in being regional intergovernmental organizations with representatives from that region's state, local, and often special-purpose governments; and they are often referred to as "councils of governments."

Analysts generally agree that prior to ISTEA, MPOs had been relatively weak decision-making bodies, exerting little influence on transportation policy (Dilger 1992). Moreover, federal funding for MPOs diminished significantly throughout the 1980s, meaning that when ISTEA passed in 1991, MPOs had smaller staffs, lower budgets, and fewer technical and administrative resources than ever before (Gage and McDowell 1995).

How MPOs Are Performing

The passage of ISTEA thus placed these obscure (and in some cases unprepared) quasi-governmental organizations at the center of efforts to translate ISTEA's holistic vision into real-world transportation programs at the regional level. The more systematic research on the performance of MPOs (U.S. GAO 1996; ACIR 1995; Lyons 1994; STPP 1996) since ISTEA's passage reveals three trends. First, ISTEA had unquestionably

Table 7.2
MPOs transportation-related responsibilities under ISTEA

Leadership	MPOs to assume the role of "leader, broker, and consensus builder" (Lyons 1994) among the many governmental and nongovernmental actors involved in metropolitan transportation planning.
Plans	MPOs to produce comprehensive transportation plans in both a short-term document (called a Transportation Improvement Plan, or TIP, which covers at least three years) and a Long Range Transportation Plan that provides a twenty-year vision for transportation investments. The TIP must list projects in order of priority, and these projects must be consistent with those in the long-range plan. Both short- and long-term plans must only include projects for which funding can reasonably be expected (called the "financial constraint" requirement).
Project selection	MPOs serving areas with populations exceeding 200,000 must select, in consultation with the state DOT, which projects to implement within their jurisdictions, except projects under the National Highway System, Bridge Replacement and Restoration Program, and the Interstate Maintenance Programs (U.S. GAO 1996).
Air quality	In CAA nonattainment areas, MPOs must make "conformity" determinations, stating that the region's transportation projects contribute to improving air quality and are consistent with state air quality plans.

boosted MPOs' transportation decisionmaking authority. Two major surveys of MPOs (U.S. GAO 1996; ACIR 1995) found, for example, that while some provisions conferred mostly symbolic authority, the "financial constraint" requirement gave MPOs real authority to determine transportation investment priorities. In some cases, the financial constraint forced MPOs to eliminate 50 percent of projects proposed by state and local transportation agencies (GAO 1996). In a second trend, however, it appears that MPOs often viewed their new authority with ambivalence, as they report being charged with too many responsibilities in relation to their resources (ACIR 1995, 44–45). And finally, due, perhaps, in part to understaffing and lack of resources, a third post-ISTEA performance trend in MPOs relates to intergovernmental conflict. MPOs report meeting resistance from other transportation policy-making agencies when attempting to assert their new authority. This seems to occur most (though not exclusively)[8] with state DOTs: MPOs generally rate their post-ISTEA relationship with state DOTs as "ineffective" or "very ineffective," and find the state DOT inertia that favors highways above other transportation modes difficult to confront (ACIR 1995; Gage and McDowell 1995).

In light of these trends, can it be said that the heightened role of MPOs in post-ISTEA transportation governance is resulting in transportation decisions that are more sustainable? It appears that advocates of the sustainable communities framework have some reason for optimism. For example, a comprehensive review of MPOs nationwide (ACIR 1997, iii) reports that "sustainable development" is a new goal "that planners of many types throughout the nation (not just transportation planners) are trying to achieve now." In addition, some analysts argue that, judged against ISTEA's holistic criteria, MPOs have proven quite effective in preventing "bad" transportation projects from going forward, although less adept in making "good" projects happen (Howitt 1997, personal interview). Both these points suggest that the sustainable communities epoch is operative in the activities of MPOs.

Most significantly, the MPOs retained the essential decision-making authority in the reauthorization of ISTEA in 1998. While the planning factors were streamlined down to seven, the essential requirement that MPOs retain financial decision-making was retained in TEA-21. Indeed, the revised planning elements can be seen as moving toward a more regionally integrated planning approach generally construed within the third epoch of the environmental movement (see chapter 1).

Open the Door and Let Them In: Toward Integrated, Participatory Transportation Planning

While a belief in "integrated, decentralized and community-based" solutions to environmental policy problems is a hallmark of the sustainable communities epoch (Mazmanian and Kraft, chapter 1), ensuring the funds to do so is another matter. As for the former, ISTEA and its successor TEA-21 have taken several steps to improve the linkages between integrated planning and investment in environmentally responsive transportation systems. The codification of this integrated approach is in a host of (now seven) planning factors that anchor metropolitan and state-level transportation planning. These planning factors, "address many of the ways that transportation relates to other . . . aspects of our society" (Braum 1994, 43) The seven factors are:

• Support the economic vitality of the metropolitan area, especially by enabling global competitiveness, productivity, and efficiency.
• Increase the safety and security of the transportation system for motorized and nonmotorized users.
• Increase the accessibility and mobility options available to people and for freight.
• Protect and enhance the environment, promote energy conservation, and improve quality of life.
• Enhance the integration and connectivity of the transportation system, across and between modes, for people and freight.
• Promote efficient system management and operation.
• Emphasize the preservation of the existing transportation system (U.S. DOT 1998).

Yet, as the discussion above of the 1962 Federal Aid Highway Act demonstrates, enshrining these planning factors in law provides no guarantee that they will be applied in practice. A crucial factor is the statute's ability to structure a potentially effective scheme of implementation (Mazmanian and Sabatier 1989). ISTEA created such a scheme: it contained two new programs—the Congestion Mitigation and Air Quality Management Program (CMAQ) and the Transportation Enhancements Program—which earmarked funds for air quality improvement and community enhancement. Moreover, ISTEA ensured more accountability to community interests by increasing the role of public involvement in the transportation planning process. These elements, largely absent in 1962, create strong financial and political incentives to actually implement ISTEA's integrated and inclusive policy approach.

During the course of ISTEA, CMAQ provided over $6 billion to transportation projects with documented air quality benefits. Priority for CMAQ funds was given for pollution-reducing "transportation control measures" (TCMs) in approved state air quality plans, and states with

Clean Air Act (CAA) nonattainment areas could not transfer CMAQ funds to other programs. By providing funds for TCMs and other projects that improve air quality, "ISTEA thus served as a funding source for CAA goals, in effect compensating for the dearth of federal aid provided by clean-air legislation" (Howitt and Altshuler 1992).[9] Beyond CMAQ's $6 billion, ISTEA's Transportation Enhancements Program set aside nearly $2.8 billion for "transportation enhancements," which were "designed to strengthen the cultural, aesthetic, or environmental aspects of transportation or to encourage greater use of non-motorized transportation" (GAO 1996, 3). ISTEA specified ten enhancement activities eligible for funding.[10]

The Enhancements Program and CMAQ present the clearest examples of how federal transportation policy targeted funds to better integrate transportation policy with broader policy objectives. Together these programs dedicated roughly $9 billion over six years to projects that improve air quality and enhance the aesthetic and cultural amenities available to communities. And these programs had demonstrable effects on the transportation system. An analysis of CMAQ's effects on air quality concluded that the program leads to annual reductions of 52,135 tons in volatile organic compounds; of 336,349 tons in carbon monoxide; and of 62,406 tons in oxides of nitrogen (EPA 1997). The Enhancements Program, moreover, gets credit for the dramatic increase in federal funds for bicycle and pedestrian facilities. During the eighteen years (1973–1990) prior to passage of ISTEA, cumulative federal spending on pedestrian and bicycle facilities totaled roughly $40 million. Yet in just four years, between 1992 and 1995, federal funding for these projects jumped to nearly $677 million (Rails-to-Trails Conservancy 1996).

As we note later, debate surrounding the reauthorization of ISTEA confirmed the difficulty in ensuring funding for innovative transportation programs. While little debate occurred surrounding the refinement of the planning factors, there was significant debate about the transportation enhancements and CMAQ funding. Indeed, many state transportation departments and road and transportation interest groups called for the elimination of both programs. Nonetheless, the resulting success in maintaining these programs due to their enormous popularity with local elected officials and community groups provides a new six-year window of opportunity to demonstrate and realize the value of a more integrated transportation investment portfolio.

Increased Public Involvement

The more integrated policy approach encouraged (and demanded) by CMAQ and the Enhancements was further buttressed by ISTEA's provisions on public involvement. As Kraft and Mazmanian note in chapter 9, an evolved public involvement process is crucial to informed local decision making. ISTEA increased public involvement in several ways:[11] State DOTs and MPOs provide for public review and comment at all "key decision points," particularly during the development and revision of major planning and programming documents. The public involvement process must be inclusive, involving diverse governmental and nongovernmental stakeholders and those "traditionally underserved by existing transportation systems." And finally, public involvement must be meaningful, as ISTEA required a process "for demonstrating explicit consideration and response to public input" during development of transportation plans and programs.

The effect of ISTEA's public involvement requirements on the transportation planning process was considerable. The sheer amount of public participation has increased, and MPOs consistently cite public involvement as the biggest change ISTEA made in their transportation planning processes (ACIR 1995). The extent to which these procedural changes have altered or improved substantive policy decisions is harder to gauge, however. One study found that MPOs believe that more public involvement produces "better" policy decisions, in that "[transportation] plans and programs are more reflective of the public's transportation needs and hence enjoy broader and stronger public support" (GAO 1996, 19). At the same time, however, MPOs tend to evaluate their own public involvement process toward the "ineffective" side,[12] and also report that narrowly focused interest groups, rather than the broader public, often dominate the public involvement process (Gage and McDowell 1995; GAO 1996).

Despite the apparently imperfect nature of post-ISTEA public involvement, ISTEA should be judged successful in bringing both a more integrated and inclusive policy approach to transportation planning.[13]

CMAQ, the Enhancements, and extensive public involvement provisions present models of the "integrated, decentralized and community-based" approach that—in the area of planning, at least—place post-ISTEA transportation policy at the forefront of the sustainable communities epoch. That these provisions remained in the TEA-21 legislation—despite the recommendation of the American Association of State Highway and Transportation Officials that public involvement be left up to the states— is testament to their perceived value by a large constituency. Yet, as the following discussion of technology suggests, the influence of ISTEA's holistic principles have not been felt equally in all dimensions of post-ISTEA transportation.

Exit Ramp to the Information Highway: New Challenges for Sustainable Transportation

Transportation, like virtually all human activities, is being profoundly affected by technological advances collectively known as the "information revolution."[14] Information technologies are now critical elements of the transportation system, performing functions such as synchronizing traffic signals, collecting highway tolls, and generating and disseminating traffic information to drivers and traffic managers. Information technologies even raise the prospect of fully automated highways, where driving involves no human sensory reaction or control (i.e., "hands-off" driving). With the post-ISTEA era focused on improving the existing highway system rather than on constructing new roads, the application of information technologies to transportation embodies former Secretary of Transportation Federico Pena's belief that "many of the [future] improvements to the transportation system will rely on the ability of private firms and public agencies to gather, process, analyze, and disseminate information" (Pena 1996, vi).

The set of technologies that will "gather, process, analyze, and disseminate" this information are known as Intelligent Transportation Systems, or ITS. Most of such systems fall into five broad categories: (1) advanced traffic management systems (ATMS), which use surveillance technologies to relay information on traffic conditions to a traffic control center; (2) advanced traveler information systems (ATIS), which provide traffic information and route guidance to drivers, transit users, and other travelers, allowing them to alter where and when to travel in response to traffic conditions; (3) advanced vehicle identification (AVI), which uses toll-tag technology to improve throughput through toll roads, as well as to enable congestion pricing; (4) commercial vehicle operations (CVO), which entails a host of wireless improvements to ease commercial trucking transactions, particularly across state lines; and (5) advanced vehicle control systems (AVCS), which seek to improve control of the automobile itself, either automatically or by assisting the driver. Fully automated highways represent the most ambitious—and, from a sustainability perspective, the most controversial—AVCS application (Diebold Institute 1995).

Federally sponsored ITS-related research dates back to the late 1960s (Chen and Ervin 1992), but not until ISTEA initiated the federal Intelligent Transportation Systems (ITS) Program were these technologies a major U.S. transportation policy priority. Between 1991 and 1995, federal spending for ITS research and development totaled roughly $828 million, and total funding through fiscal year 1997 exceeded $1 billion (Congressional Budget Office 1995; Horan, Hempel, and Bowers 1995). And in early 1996, the U.S. DOT launched "Operation Timesaver," with the goal of deploying an "Intelligent Transportation Infrastructure" in seventy-five of the largest U.S. metropolitan areas within the next ten years. Supplementing these federal activities are many state and local government ITS programs, and private sector ITS investment is growing and predicted to be the primary source of ITS funding.[15]

That an aggressive ITS program arrived as part of ISTEA raises important policy questions. Will ITS—conceived and designed by transportation engineers largely in the pre-ISTEA era—reflect the more traditional emphasis on moving traffic as quickly as possible? Or will principles of the sustainability epoch serve to guide ITS investments? At the outset of ISTEA, opinion among environmental analysts was decidedly mixed. Some cautiously optimistic analysts argue that, properly directed, "ITS could be the most important enabling technology driver in decades for reform and progress in American transportation, winning for our citizens sustainable high-wage jobs, reduced traffic delay, more livable communities, and a healthy environment" (Replogle 1995, 59). By contrast, others view ITS unequivocally as an environmental and social Pandora's box. Despite promising a more "efficient" transportation system, such analysts often argue that "A principal objective of [ITS]—to minimize total vehicle-hours of delay—has little in common with the social imperative of reducing the environmental impacts of driving" (Gordon 1992, 24–25).

The experience during the ISTEA period (1991–1998) sugggested that the integrated approach advocated in the third environmental epoch was, with notable exceptions, reluctantly being followed by many in the ITS community (CBO 1995; Public Technology, Inc., 1998). While many states incorporated some aspect of travel-reducing elements into their ITS deployment plans, overall priority and investment for environmentally beneficial actions remained relatively low (CBO 1995). Moreover, much of ITS decision making remained outside traditional channels of transportation investments, allowing for only modest levels of public dialogue and debate. Several steps were taken, including our own research, to outline how ITS systems could be deployed consistent with sustainability parameters, and, in selected cases, there was movement in this direction (Humphrey Institute 1998; PTI 1998; Shaheen et al. 1998).

Conceptual Links between ITS and Holistic Transportation
The potential link between ITS and sustainable transportation stems from ITS's ability to create a transportation system rich in information, or what might be called an "information-intensive" transportation system. Such a system raises two prospects. First, it means using information instead of new lanes, roads, and highways as a way to increase the capacity of the transportation system. In this sense, ITS "substitutes information for stuff," resulting in capacity enhancements that use fewer material resources, consume less open space, and reduce the noise and community disruption related to new roads. ITS thus supports an underlying premise of "sustainability thinking": that the Earth's resource base has limits, that some of those limits are being approached, and, that, therefore, sustainable development depends on accommodating economic growth while consuming fewer resources.[16]

Beyond potentially substituting for physical elements of the transportation system, the information that ITS provides may also enhance the system's performance. Critical in the ISTEA era, however, is that "enhanced performance" be defined broadly to include greater traffic efficiency and a reduction in the transportation system's negative externalities. ITS can contribute to this broader notion of enhanced performance by providing information that allows for greater operational control of the transportation system. Achieving more control of the system, in turn, increases the opportunities to address specific purposes, including broad social, economic, and environmental goals.

Table 7.3 illustrates select ITS applications that facilitate greater control of the transportation system by channeling information to system managers and users.[17] "Remote sensing,"[18] for example, can generate emissions data and assist air quality officials in targeting "gross polluters." Another example is "congestion pricing," or charging drivers a fee that varies with the level of traffic on a roadway. Congestion pricing conveys information (in the form of price signals) that alerts drivers to the overall social and environmental costs of driving, making them aware that driving imposes external costs while encouraging more environmentally benign travel behavior.[19]

The Real World of ITS Deployment

Drawing conceptual links between ITS and principles of the sustainable communities epoch is one thing; making those links is another. Such links appear to be occurring in some cases: Minnesota's Department of Transportation, for example, has initiated a Sustainable Transportation Initiative (STI) to implement ITS programs consistent with sustainability principles (Minnesota Department of Transportation 1998). The federal ITS program—having symbolically embraced a more holistic, less highway-focused approach by changing its name in 1994 from "Intelligent Vehicle Highway Systems" to "Intelligent Transportation Systems"—lists environmental quality as one of its primary goals (Hennessey and Horan 1995).

Yet federal spending priorities tell a somewhat different story. According to a Congressional Budget Office (CBO 1995) report, only $5.6 million—totaling 1.2 percent of federal ITS funds obligated through 1994—went to projects in which environmental concerns were the primary motive. This compares with $304.6 million (65.3 percent of federal funding obligated through 1994) spent on travel and traffic management projects. These data led the CBO to conclude that:

Table 7.3
ITS-generated information and sustainability

ITS category	Application	Flow of information	Contribution to sustainability
Advanced traffic management	Traffic signal synchronization	Traffic information to traffic managers allows retiming of signals to optimize traffic flow	Reduces energy usage and emissions related to "stop" and "go" traffic and congestion
	Incident detection	Incident (i.e., freeway accident) information to traffic managers allows faster emergency response, retiming of ramp meters, etc.	reduces energy usage and congestion-related emissions
Traveler information	Pre-trip traveler information	Traffic information to traveler allows for shift in travel time, route, or mode	Reduces energy usage, congestion-related emissions and/or the number of trips/single occupant vehicles
	En-route traveler information	Traffic information to driver allows shift in route	Reduces energy usage congestion-related emissions
Advanced vehicle identification	Congestion-sensitive road tolls (i.e., congestion pricing)	Information to drivers (in the form of price signals) that relays full social and environmental costs of driving	Reduces energy usage and emissions by reducing number of trips/single occupant vehicles, reducing congestion, and perhaps encouraging less auto-dependent land-use patterns (i.e., less sprawl)
Other	Remote sensing of emissions	Vehicle emissions information to drivers and/or air quality managers	Aid in targeting "gross polluters" (10% of vehicles responsible for roughly 50% of emissions)
	Demand-responsive transit services	Information to transit managers and transit riders on supply/demand status related to transit	Reduces emissions and energy usage by encouraging use of transit; helps create more equitable distribution of transportation services to underserved populations (i.e., handicapped, elderly)

Among the objectives [for the ITS program] set forth by the Congress, the one that seems to have received the least attention is the environment. Although some of the travel management projects could benefit the environment, how they might do so is not entirely clear because short-term reductions in traffic and congestion could lead to greater numbers of vehicles on the road, resulting in even greater pollution. (p. 44)

The CBO report rightly points out that funding for ITS projects does not necessarily represent a zero-sum game between promoting environmental goals versus those related to mobility. As TEA-21 reauthorized the ITS program at $1.3 billion, ensuring enhanced linkage between ITS systems and the broader goals of the post-TEA-21 era remains a major area of transportation policy attention (DOT 1998). More broadly, while ISTEA and now TEA-21 have provided an overall framework for moving toward more sustainable transportation systems, individual elements within the legislation—such as ITS—each have their own track record on the extent and pace with which these broader principles are adopted.

Arriving Home: Summary and Conclusions

The reauthorization of ISTEA in 1998 provided a test of the political value of sustainability principles. At the outset, the principle had political value to the administration. In introducing the legislation, President Clinton stated:

We're building a bridge to sustainable communities that can last and grow and bring people together over the long run . . . For too long, too many people have believed that strong transportation and a clean environment could not go hand in hand. This bill proves that that is not true . . . Make no mistake about it, this is one of the most important pieces of environmental legislation that will be considered by the Congress in the next two years. (Clinton 1997)

Clinton's statement highlights both the prospects and problems facing holistic transportation that were highlighted throughout this chapter. Encouraging to "sustainability" advocates is Clinton's nomenclature: his prominent use of the term "sustainable communities" marks the term's arrival into the mainstream national transportation policy debate. Moreover, Clinton makes clear his view that ISTEA reauthorization represents critical environmental legislation.[20] This suggests that a central notion of the sustainable communities framework—that transportation policy be viewed in an integrative fashion dictated by broader social and environmental objectives—now suffuses conventional thinking about the role of transportation in society.

Yet Clinton's remarks (as well as our effort to characterization them) underscore perhaps the most vexing problem facing proponents of sustainable transportation, namely, that policy prescriptions that invoke "sustainability" are often overly general and almost poetic in tone. This problem persists, we believe, because it is easier to identify what holistic transportation is not (i.e., automatically building more roads as a knee-jerk policy response to urban traffic congestion) than to describe what it is. Richard Mudge (1994) states the problem as such:

Around the turn of the century, the clear goal [of transportation policy] was to get the farmer out of the mud. From 1956 on, the clear goal was to build the interstate system. There is no real goal now.

No real goal, but instead multiple goals that are not only subtle, often vague, and at times irreconcilable, but also that vary greatly across and within communities. The consequences of a lack of political consensus became clear as different interest groups began to stake out positions in the reauthorization debate. Four distinct approaches emerged in 1996, as Congress began to hold hearings prior to a planned 1997 markup of the legislation. First to weigh in were the "donor" states, whose proposed STEP-21, focused upon attaining two objectives: getting a larger share of formula funds back to their states, and ensuring that these funds could be spent as flexibly as possible. The STEP-21 proposal which was eventually endorsed by nineteen state departments of transportation, thus called for elimination of the CMAQ and enhancement funding programs.

The road-building and automotive industries came forward through the "Keep America Moving Coalition," which promoted a "Highways Only Transportation Efficiency Act" (HOTEA). That would have abolished the federal transit program, the Enhancements Program, and the CMAQ program, and would have required 80 percent of all funds to be spent on road construction. The American Highway Users Alliance and the American Automobile Association mounted fierce campaigns against the Enhancements Program, alleging that the "diversion" of highway funds to projects like bicycle trails and historic preservations was resulting in a loss of lives, as the funds could have been spent otherwise on highway safety projects. The third proposal of significant import was advanced by Representative John Kasich (R-Ohio) and Senator Connie Mack (R-Florida). They called for massive devolution of transportation programs to the states, with only a small Interstate and research program at the federal level and the balance of federal gas taxes being repealed. Their proposal got little support when it became clear that state legislatures and governors would have to raise gas taxes at the state level to make up for the loss of federal revenues. This program would also, of course, have eliminated the CMAQ and Enhancements Program.

The final major proposal was the "ISTEA Works," supported by environmental and local government interest groups and by twenty-three state departments of transportation. ISTEA Works proposed to keep ISTEA's basic program structure and planning framework intact.

The fact that three out of the four major reauthorization proposals would have abolished the enhancement and CMAQ programs was a clear indication that the traditional interest groups in transportation were threatened by federal legislation that opened up issues of the sustainability of our transportation system. It soon became clear, however, that the programs were exceedingly popular, and the strong support of the U.S. Conference of Mayors, the National League of Cities, and the National Association of Counties for these programs, along with the active advocacy of all the major environmental groups, soon made their preservation a foregone conclusion. Transportation interests then tried to amend the programs out of existence by allowing states to transfer CMAQ and Enhancements funding to road projects and by defining road widening projects as an acceptable air quality strategy.

In July 1998, when President Clinton signed TEA-21 into law, it contained few of these proposals to roll back ISTEA's reforms. Instead it increased Enhancements and CMAQ funding by over 40 percent, added a new "transit enhancements" category, and preserved ISTEA's planning framework. In addition, the new law extended ISTEA's definition of sustainability by adding two new programs: a Job Access Program, which dedicated $150 million to helping with transportation aspects of welfare reform, and a new Transportation Community and System Preservation pilot program, which funds regions seeking to integrate transportation, land use, and sustainability. TEA-21 thus advanced rather than set back the idea that federal policy could provide a framework for community efforts to achieve sustainability.

That a reasonable course emerged from these debates suggests that federal parameters play a continuing role in establishing the parameters for the third epoch of environmental policy. Yet, as Kraft and Mazmanian outline in chapter 9, the emergent context is at the local and regional level. The linkages (and tensions) between federal and local desires became manifest during the TEA-21 debates, and will continue through the course of this legislation. MPOs, for example, face tremendous challenges in the nation's rapidly growing areas, such as "Sunbelt" regions of the South and Southwest that have embarked on aggressive capital expansion (i.e., highway construction) programs to accommodate this growth (Howitt, personal interview). And, as outlined in the ITS section, regions throughout the United States will confront the challenge of integrating these rapidly advancing technologies with the broader planning mandates of ISTEA and its successor, TEA-21.

Are such programs, which are often politically popular and deemed necessary to preserve a community's "quality of life," consistent with ISTEA's holistic approach? With respect to planning and public involvement, are "public interest advocates" or others—in the name of "sustainability," a "holistic approach," or the "community interest"—merely using ISTEA to pursue their particular vision(s) of the ideal community? And finally, ITS's theoretical links to sustainable transportation notwithstanding, the technologies in fact grow out of a traffic engineering community concerned most with moving vehicles. Is it any surprise, therefore, that efforts to apply the technologies to broader applications have been decidedly mixed? ITS thus remains a promising but somewhat lagging aspect of the postinterstate philosophy.

The ambiguity surrounding the constituent elements of holistic transportation remains a continual source of frustration for transportation policy makers. Yet, paradoxically, the nuanced, context-specific nature of ISTEA's policy goals may also be the law's underlying strength. This is because transportation strategies that are consistent with sustainable communities are, by definition, contingent on the needs and desires of particular communities. Since these needs and desires vary from place to place, change over time, and are subject to ongoing debate, sustainable transportation entails a continual process of tailoring transportation strategies to specific communities and places (Dittmar 1995). Indeed, this is ISTEA's hallmark: an attempt to reconnect transportation policy with the specific attributes of "place"—the unique, multifaceted areas where people live, work, shop, and participate in community life. "One-size-fits-all" transportation solutions ignore the particularities of place and are therefore inimical to holistic notions of transportation.

A new pilot program in TEA-21 reflects this context-based view of sustainable places. Known as the "Transportation and Community and System Preservation Pilot Program," this program will provide a series of grants to local communities, the aim being the demonstration of integrated transportation and community programs (DOT 1998). While modest in scale (under $1 billion), it represents an important recognition of this new approach to designing transportation projects. And consistent with parameters outlined by Kraft and Mazmanian in chapter 9, it includes research and testing of integrated environmental approaches as means to enhance the empirical base for this new model. The initial solicitation for project applications under this new program closed in November 1998, with more than 600 communities vying for only about $20 million in first-year funds.

As the empirical base from this program and other TEA-21 programs grows, transportation legislation appears ready to embrace a holistic connection between this infrastructure and the environment. However, it cannot guarantee that this occurs throughout the country. In this light, federal transportation legislation has served as a point of departure for integrated environmental policy. The point of arrival will be at the local-community level as the ambitions of national legislation meet the context of local priorities, values, and preferences.

Notes

The authors acknowledge the support of the U.S. Department of Transportation (DOT) for conduct of background research on which some of this chapter is based. The authors also express their appreciation for the support of Lee Munnich, Jr., Humphrey Institute, University of Minnesota, for his leadership on the DOT-sponsored Transportation Technologies for Sustainable Communities Study. Portions of this chapter have appeared in Thomas A. Horan and Daniel

Jordan, "Institutional and Operational Challenges," chapter 2 in *Wired to Go: Final Report on Transportation Technologies for Sustainable Communities,* Minneapolis, MN: Humphrey Institute, 1998.

1. Moynihan's critique of the Interstate Highway Program would be echoed by numerous analysts in subsequent years. Another widely circulated early critique of the Interstate Program is found in Helen Leavitt, *Superhighway-Superhoax* (1970). For a critical contemporary account of the Interstate Highway Program, see Lewis (1996).

2. "Holistic" is defined in the *American Heritage Dictionary of the English Language,* Third Edition, (Houghton Mifflin 1992) as "a. Emphasizing the importance of the whole and the interdependence of its parts. b. Concerned with the wholes rather than analysis or separation into parts." The terms "holistic," "sustainable transportation," "sustainable communities," "sustainability," and "livable communities" will be used interchangeably throughout this chapter to describe attempts at balancing multiple social, economic, and environmental goals to improve the quality of life in communities.

3. All Highway Trust Fund monies were dedicated to highways until 1973, when urban mass transit programs became eligible for a portion of the funds (Weiner 1987).

4. The beginning of the so-called freeway revolts, in which citizen protests succeeded in delaying or canceling freeway construction projects around the nation, occurred in 1959, over the proposed Embarcadero Freeway in San Francisco. Other early anti-Interstate activity occurred in Boston, Massachusetts, Oakland, California, and New Orleans, Louisiana. See Leavitt 1970, 53–109; and Lewis 1997.

5. An EIS is a "detailed statement" that discusses "the environmental impacts of the proposed action, any adverse environmental affects which cannot be avoided, alternatives to the proposed action, the relationship between local short-term uses of man's environment and the maintenance and enhancement of long-term productivity, and any irreversible and irretrievable commitment of resources which would be involved should the proposed action be implemented" (Council on Environmental Quality 1994, 349).

6. EPA subsequently waived these deadlines several times (see Kraft 1996, 86).

7. "Conformity" refers to the relationship between state transportation plans and state air quality plans. "Simply put, conformity to the State Implementation Plan (SIP) for air quality attainment means that the transportation plans and programs for the non-attainment region will not, 1) cause any new violations of National Ambient Air Quality Standards (NAAQS), 2) cause any worsening of existing violations, and 3) delay the region's effort to attain the NAAQS in a timely manner" (Siwek 1994, 81).

The significance of the 1990 CAAA should not be underestimated. Pas (1995, 58) suggests "that the CAAA, through the increased responsibility that is assigned to mobile source emissions for improving the air in the nation's cities and the related analysis and policy requirements, will be seen as the piece of legislation that really changed the face of urban transportation planning in the United States."

8. State DOTs are not the sole source of conflict for MPOs, however. According to a *Los Angeles Times* article, a recent transportation plan developed by Southern California's MPO (the Southern California Association of Governments, or SCAG) was sharply criticized by public officials in Orange County (one of the counties within the SCAG region). The article stated that the SCAG plan "failed to include a third of Orange County's priority transportation projects," and prompted County Supervisor Jim Silva to state that "Orange County has to make decisions for Orange County." Several county supervisors, the article said, "expressed interest in seeking federal legislation that would give a county agency more authority in funding road projects." See Shelby Grad. "Regional Road Plan Attacked by Supervisors." *Los Angeles Times,* Orange County edition, March 19, 1997, B1.

9. Even Denno (1994, 277), who believes most analyses overstate the degree to which ISTEA deviated from past federal transportation policies, nonetheless highlights the new direction in CMAQ: "There was no program even distantly resembling CMAQ prior to ISTEA . . . This program represents a significant departure from past federal programs. It indicates a recognition of the fact that a federal transportation program can be designed to achieve transportation as well as other social objectives."

10. The enhancement activities listed are: (1) provision of bicycle and pedestrian facilities; (2) acquisition of scenic easements and scenic or historic sites; (3) scenic and historic highway programs; (4) landscaping and other scenic beautification; (5) historic preservation; (6) rehabilitation and operation of historic transportation buildings, structures, and facilities; (7) preservation of abandoned transportation corridors; (8) archeological planning and research; (9)control and removal of outdoor advertising; (10) mitigation of water pollution caused by highway runoff.

11. This and the other referenced public involvement provisions are found in ISTEA's implementing rules, 23 CFR 450.212.

12. Gage and Mcdowell asked 181 MPO officials to "please rate the overall effectiveness of public involvement as it relates to local policymaking on a ten point scale from 1 (very ineffective) to 10 (very effective)." The average rating was 4.68. Interestingly, the responses differed by type of MPO: "Large MPOs in CAAA non-attainment areas" had the highest average rating (5.90), while "large MPOs in attainment areas" had the lowest rating (3.33).

13. ISTEA's significance in this area is reflected in the favorable grading the law's implementation received by one of the nation's most prominent advocates of sustainable transportation. The Surface Transportation Policy Project, a nonprofit coalition of some 175 groups that advocate for environmentally and socially sensitive transportation policies concluded that "the legislation has transformed the transportation planning process . . . by recognizing the long-term economic, social, and environmental impacts of transportation decisions."

14. By 1980, the information sector accounted for nearly half the U.S. labor force (Beniger 1986). Numerous popular depictions of the shift to an information society have reached mass audiences, most notably Alvin Toffler's *The Third Wave.*

15. The National ITS Program Plan (ITS America 1995) estimates that the private sector will ultimately pay 80 percent of the costs of all ITS investments. Private-sector investment in ITS has focused to date on traffic information products and services, especially on navigational aids.

16. We take the phrase "substituting information for stuff" from Robert B. Shapiro, chairman and CEO of Monsanto Company. In a 1997 interview published in a *Harvard Business Review* article entitled "Growth Through Global Sustainability," he underscored the indispensable role of information in promoting sustainable development: Using information is one of the ways to increase productivity without abusing nature . . . A closed system like the earth's can't withstand a systematic increase of material things, but it can support exponential increases of information and knowledge. Sustainability and development might be compatible if you could create value and satisfy people's needs by increasing the information component of what's produced and diminishing the amount of stuff" (p. 882).

17. For an illustrative overview of ITS applications to sustainability, see Public Technology, Inc., "The Road Less Travelled: Intelligent Transportation Systems for Sustainable Communities," Washington, D.C., Public Technology, Inc. 1998.

18. "Remote sensing" refers to technologies that can measure the exhaust emissions from vehicles as they pass a roadside detector.

19. One study of congestion pricing, for example, found that fees of between 10 cents and fifteen cents per mile could reduce travel during that period by 10 percent to 15 percent (National Research Council 1994), thus reducing congestion-related emissions and perhaps leading to a net reduction in automobile use.

20. In his announcement of the Senate's version of ISTEA reauthorization, Senator Robert Chafee (R-Rhode Island) also reaffirmed the fundamental elements of ISTEA.

References

Beniger, James R. 1986. *The Control Revolution: Technological and Economic Origins of the Information Society.* Cambridge: Harvard University Press.

Braum, Phil. 1994. "ISTEA Planning Factors in the Transportation Planning Process." In *ISTEA Planners Workbook*, ed. Margaret Franko. Washington DC: Surface Transportation Policy Project.

Cervero, Robert. 1995. "Why Go Anywhere?" *Scientific American* (September): 92–93.

Cervero, Robert. 1989. *America's Suburban Centers: The Land Use-Transportation Link.* Boston: Unwin Hyman.

Chen, K., and Ervin, R.D. 1992. "Worldwide IVHS Activities: A Comparative Overview." In *Vehicle Electronics Meeting Society's Needs: Energy, Environment, Safety.* Society of Automotive Engineers, 339–349.

Clinton, William. 1997. *Press Release on NEXTEA,* Washington, DC: The White House, March 12.

Council on Environmental Quality. 1994. *Environmental Quality: 24th Annual Report of the Council on Environmental Quality.* Washington, DC: Council on Environmental Quality.

Davidoff, Paul. 1965. "Advocacy and Pluralism in Planning." *Journal of the American Institute of Planners* 31(4): 331–338.

Denno, Neal. 1994. "ISTEA's Innovative Funding: Something Old, New and Borrowed." *Transportation Research Quarterly* 24, 3: 275–285.

Diamond, Henry L., and Patrick F. Noonan, eds. 1996. *Land Use in America.* Washington, DC: Island Press.

Diebold Institute for Public Policy Studies, Inc. 1995. *Transportation Infostructures: The Development of Intelligent Transportation Systems.* Westport, Connecticut: Praeger Publishers.

Dilger, Robert J. 1992. "ISTEA: A New Direction for Transportation Policy." *Publius, The Journal of Federalism* 22, 3 (Summer): 67–78.

Dittmar, Hank. 1995."A Broader Context for Transportation Planning: Not Just an End in Itself." *Journal of the American Planning Association* 61, 1 (Winter): 7–13.

Gage, Robert W., and Bruce D. McDowell. 1995. "ISTEA and the Role of MPOs in the New Transportation Environment: A Midterm Assessment." *Publius, The Journal of Federalism* 25, 3 (Summer): 133–154.

Gifford, Jonathon L., Thomas A. Horan, and Louise G. White. 1994. "Dynamics of Policy Change: Reflections on 1991 Federal Transportation Legislation." *Transportation Research Record 1466* (December): 8–13.

Giuliano, Genevieve, and Martin Wachs. 1992. "Managing Transportation Demand: Markets versus Mandates." In *Congestion Pricing for Southern California: Using Market Pricing to Reduce Congestion and Emissions.* Los Angeles: The Reason Foundation, September.

Gordon, Deborah. 1992. "Intelligent Vehicle/Highway Systems: An Environmental Perspective." In *Transportation, Information Technology and Public Policy: Institutional and Environmental Issues in IVHS, Proceedings: A Workshop on Institutional and Environmental Issues, Asilomar Conference Center, Monterey, California, 26–28 April 1992.* eds Jonathon L. Gifford, Thomas A. Horan, and Daniel Sperling. Arlington, VA: The Institute of Public Policy, George Mason University. Pp. 9–27.

Hamilton, Alexander, James Madison, and John Jay. *The Federalist Papers.* England: NAL Penguin Inc.

Hennessey, Thomas, and Thomas Horan. 1995. *National Conference on Itelligent Transportation Systems and the Environment: Conference Proceedings.* Fairfax, VA: George Mason University.

Horan, Thomas A., Lamont C. Hempel, and Margo Bowers. 1995. *Institutional Challenges to the Development and Deployment of ITS/ATS Systems in California. California PATH Research Report,* UCB-ITS-PRR-95-17 (May). Berkeley: California PATH.

Howiit, Arnold. 1997. Personal interview with Thomas Horan, Cambridge, MA. July 17.

Howitt, Arnold M., and Alan Altshuler. 1992. *The Challenges of Transportation and Clean Air Goals.* Unpublished Executive Training Materials, Alfred Taubman Center for State and Local Government, John F Kennedy School of Government, Harvard University, October.

Humphrey Institute, 1998. *Wired to Go: Final Report on Transportation Technology for Sustainable Communities.* Minneapolis, MN: University of Minnesota.

Kraft, Michael E. 1996. *Environmental Policy and Politics: Toward the Twenty-First Century.* New York: HarperCollins.

Leavitt, Helen. 1970. *Superhighway-Superhoax.* New York: Doubleday and Company.

Lewis, Peirce. 1996. "The Landscape of Mobility." In *The National Road,* ed. Karl Raitz. Baltimore: The Johns Hopkins University Press. Pp. 3–44.

Lewis, Thomas. 1997. *Divided Highways: Building the Interstate Highways, Transforming American Life.* New York: Viking Press.

Lyons, William. 1994. *The FTA-FHWA MPO Reviews—Planning Practice Under ISTEA and the CAAA.* John A. Volpe Transportation Systems Center, U.S. Department of Transportation, January.

Mazmanian, Daniel, and Paul Sabatier. 1989. *Implementation and Public Policy,* with a new postscript. Lanham, MD: University Press of America.

Minnesota Department of Transportation. 1998. Sustainable Transportation Business Plan, St. Paul, MN: Minnesota Department of Transportation.

Morehouse, Thomas A. 1971. "Artful Interpretation: The 1962 Highway Act." In *Metropolitan Politics: A Reader.* 2nd ed., ed. Michael N. Danielson. Boston: Pp. 353–368.

Moynihan, Daniel Patrick. 1991. Quoted in Mike Mills, "Highway and Transit Overhaul Is Cleared for President." *Congressional Quarterly Weekly Report,* November 30, 3518–3522.

Moynihan, Daniel Patrick. 1960. "New Roads and Urban Chaos." *The Reporter* 22, 8: 13–20.

Mudge, Richard. 1994. Quoted in Margaret Franko, ed, *ISTEA Planners Workbook,* Washington, DC: Surplus Transportation Policy Project.

Munro, David A. 1995. "Sustainability: Rhetoric or Reality?" In *Sustainable World: Defining and Measuring Sustainable Development,* ed. Thaddeus C. Tryzyna. Sacramento, CA: International Center for the Environment and Public Policy and The World Conservation Union. Pp. 27–35.

National Research Council, Transportation Research Board. 1994. *Curbing Gridlock: Peak-Period Fees to Relieve Traffic Congestion.* Special Report 242. Washington, DC: National Academy of Sciences.

Pas, Eric I. 1995. "The Urban Transportation Planning Process." In *The Geography of Urban Transportation,* 2nd edition, ed. Susan Hanson. New York: Guilford Press.

Pena, Federico. 1996. Remarks Prepared for Delivery at the Transportation Research Board Annual Meeting. Washington, DC, January 10.

Public Technology, Inc. 1998. *Roads Less Traveled: Intelligent Transportation Systems for Sustainable Communities,* Washington, DC: Public Technology Inc.

Rails to Trails Conservancy. 1996. "Transportation Enhancement Financial Summary." Washington, D.C.: Rails to Trails Conservancy.

Replogle, Michael. 1995. "Intelligent Transportation Systems for Sustainable Communities." In *National Conference on Intelligent Transportation Systems and the Environment: Conference Proceedings,* ed. J. Thomas Hennessey and Thomas A. Horan. Arlington, VA: Institute of Public Policy, George Mason University. Pp. 53–59.

Rose, Mark H. 1990. *Interstate: Express Highway Politics, 1939–1989.* Revised ed. Knoxville, TN: The University of Tennessee Press/Knoxville.

Shaheen, Susan, Troy Young, Daniel Sperling, Daniel Jordan, and Thomas Horan, 1998. *Identification and Prioritization of Environmentally Beneficial Intelligent Transportation Technologies.* Report prepared for California PATH Program, Berkeley, CA: California PATH.

Shapiro, Robert B. 1997. Quoted in Joan Magretta, "Growth through Global Sustainability." *Harvard Business Review* (January–February): 79–88.

Siwek, Sarah. 1994. "Conformity." In *ISTEA Planner's Workbook.* ed. Margaret Franko. Washington, DC: Surface transportation Policy Project, October.

U.S. Advisory Commission on Intergovernmental Relations. 1997 (February). *Planning Progress: Addressing ISTEA Requirements in Metropolitan Planning Areas.* Washington, DC: U.S. Advisory Commission on Intergovernmental Relations.

U.S. Advisory Commission on Intergovernmental Relations. 1995. *MPO Capacity: Improving the Capacity of Metropolitan Planning Organizations to Help Implement National Transportation Policies.* (May) Washington, DC: U.S. Advisory Commission on Intergovernmental Relations.

U.S. Congress. Congressional Budget Office. 1995 (October). *High-Tech Highways: Intelligent Transportation Systems and Policy.* Washington, DC: Congressional Budget Office.

U.S. Department of Transportation, 1998. *Overview of TEA-21 Provisions,* Washington, DC: GPO.

U.S. Department of Transportation, Federal Highway Administration. 1992. *Our Nation's Highways: Selected Facts and Tables.* FHWA-PL-92-004. Washington, DC: GPO.

U.S. Environmental Protection Agency, 1997. *The Emission Reduction Potential of the Congestion Mitigation and Air Quality Program.* Washington, DC: EPA, May 17.

U.S. Environmental Protection Agency and U.S. Department of Transportation. 1993. *Clean Air through Transportation: Challenges in Meeting National Air Quality Standards.* Washington, DC: GPO, June.

U.S. General Accounting Office. 1996. *Urban Transportation: Metropolitan Planning Organizations' Efforts to Meet Federal Planning Requirements.* GAO/RCED-96-200.

Webster, Frank, and Kevin Robins. 1986. *Information Technology: A Luddite Analysis.* Norwood NJ: Ablex Publishing Corporation.

Weiner, Edward. 1987. *Urban Transportation Planning in the United States: An Historical Overview.* New York: Praeger Publishers.

8

Sustainability in a Regional Context: The Case of the Great Lakes Basin

Barry G. Rabe

The Great Lakes Basin may constitute the ultimate test of a North American region's capacity to adhere to the tenets of sustainable development. Covering a surface area of over 300,000 square miles, the basin holds one-fourth of the world's freshwater supply and is the only glacial feature on the Earth's surface that is visible from the moon. The challenge of forging sustainable strategies for the basin stems not only from its imposing physical scope but also from the intense concentration of population and industrial activity within its boundaries for more than a century. One-fourth of the entire population of Canada resides there, and nearly one-half of all Canadian manufacturing occurs within the basin. In turn, one out of every nine Americans lives in the basin, and nearly one-fourth of total American manufacturing takes place in the region (see table 1). Virtually all standard measures of toxic pollutants and hazardous wastes generated in both nations indicate that provinces and states in the basin are national leaders in the volumes of such contaminants generated and released.

The environmental sustainability of the basin first began to be seen as highly jeopardized about three decades ago, at the outset of the modern environmental movement. The seeming ecological demise of Lake Erie, a series of blazes along waterways such as the Cuyahoga River, and mounting evidence that other lakes (particularly Michigan and Ontario) were seriously threatened served to trigger considerable debate over whether ecological health could be regained, or if the basin was trapped in an irreversible downhill slide. In addition to conventional contamination threats, the basin also appears particularly vulnerable to the effects of cross-media pollution. Large surface waters, long residence time for

Table 8.1
Population of the Great Lakes Basin

Jurisdiction	Lake Superior	Lake Huron	Lake Michigan	Lake Erie	Lake Ontario	Basin totals for jurisdictions
Ontario	181,573	1,191,467	Not applicable	1,664,639	5,446,611	8,487,210
Indiana			1,087,494	339,264		1,426,758
Illinois			3,494,115			3,494,115
Michigan	142,606	1,502,687	3,007,954	4,646,843		9,300,090
Minnesota	212,796					212,796
New York				765,537	2,702,065	3,467,602
Ohio				4,023,625		4,023,625
Pennsylvania				242,261	2,219	244,480
Wisconsin	70,416		2,467,463			2,537,609
United States total*	425,548	1,502,687	10,057,026	10,017,530	2,704,284	24,707,075
Canada total**	181,573	1,191,467	Not applicable	1,664,639	5,446,611	8,487,210
Great Lakes Basin total	607,121	2,694,151	10,057,026	11,682,169	8,150,895	33,384,157

* United States total is based on 1990 census data.
** Canada total is based on 1991 census data
Sources: Environment Canada and United States Environmental Protection Agency.

water before it circulates out of the basin, and prevailing air currents make all five lakes vulnerable to air deposition. More than 1,000 organic compounds and heavy metals are detectable in all five lakes of the basin, with air the only conceivable source in numerous instances (Colborn 1990, ch. 4). Other pervasive sources of cross-media pollutant transfer that often defy jurisdictional boundaries and threaten the basin include groundwater discharge, landfill leaching, pesticide and topsoil runoff from agricultural activity and residential development, and release from contaminated lake-bottom sediments.

The daunting scope of such problems has served to focus growing interest on ways to begin to view the basin as a unified ecosystem and to assure long-term sustainability. A flurry of collective activity at multiple levels of government has occurred in the basin over the past quarter-century, much of it linked by the common goal of comprehensive basin protection. Some outright improvements in basin environmental quality have been registered, most notably in nutrients such as phosphorus, but also in some areas of toxic release and contamination involving all media. Stabilization of environmental quality appears to have occurred in other areas, despite continued growth of population and industrial activity. Virtually every unit of government in the basin has played some role in this process. While very loosely coordinated, these numerous strategies and programmatic efforts have contributed to an unprecedented degree of unity and common vision for the basin as a collective whole.

This evolution defies, in many respects, the conventional depictions of how regional ecosystems are most likely to be protected. On the one hand, many analysts contend that only central, hegemonic oversight can be relied on to secure the consent of multiple stakeholders in protection of common resources. Drawing on Garrett Hardin's classic work on the plundering of the commons, Robert Heilbroner, William Ophuls, A. Stephen Boyan, Jr., and others have offered forceful assertions that only strong, top-down authority can foster long-term ecological viability (Ophuls and Boyan 1992). In the Great Lakes, periodic proposals for comprehensive regional "superagencies" with vast powers over development and environmental regulation have followed such lines of analysis (Caldwell 1988, 325–1927). Lester Milbrath, for example, has endorsed creation of a Great Lakes Futures Board, with overarching responsibility for a "total" environmental impact assessment process that would govern all major development activity within the basin. Such a board, in Milbrath's view, would be joined by a Great Lakes Court possessing extensive cross-jurisdictional authority over resolution of environmentally related disputes (Caldwell, 1988, 150–154). In turn, some analysts of the Great Lakes Basin have tended in recent years to attribute most positive developments to the actions of existing central institutions, such as the International Joint Commission (IJC) and its respective Great Lakes Water Quality Agreements (Caldwell 1988; Durnil 1995). Overall, a good deal of published policy analysis on the basin in recent decades is clearly inclined toward more centralized strategies.

This top-down view is countered vigorously by a competing school of thought that perceives bottom-up initiative and partnership as driving forces behind any move toward sustainability. Extremely skeptical of the capacity of central institutions to secure broad stakeholder support for implementable approaches to sustainability, such analysts would tend to attribute any positive developments in the basin to arrangements established independently of binational or central government authority. DeWitt John contends that "civic environmentalism" is thriving in American states and localities, unleashing creative approaches to complex environmental issues and filling voids left by prior command-and-control regulatory initiatives from Washington, D.C. (John 1994). Elinor Ostrom and colleagues examining the protection of "common-pool resources," such as river basins, essentially concur with John, and emphasize "common design principles" that facilitate cooperative and effective protection of potentially vulnerable ecosystems. These principles generally find little if any constructive role for central institutions to play, with creative arrangements for collective choice cultivated through locally based networks (Ostrom 1990; Ostrom, Gardner, and Walker 1994). Applied to the Great Lakes Basin, such a view would contend that any movement toward sustainability in recent decades can be attributed primarily to local stakeholders working creatively and constructively toward a shared governance system for resource protection.

This analysis of the recent evolution of the Great Lakes Basin acknowledges some merit to the claims of both rival camps, but submits that any shift toward regional sustainability stems from a multiplicity of factors that defy either a top-down or bottom-up categorization. Instead, a multi-institutional system of regional governance has evolved that attempts to replace conventional, medium-based pollution control efforts with alternative approaches designed to minimize cross-media and cross-boundary pollutant transfer, and prevent generation of new contamination. This system draws on significant contributions from binational entities such as the IJC, respective federal, state, provincial, and local governments, and multistate and multiprovince confederations. More than 650 stakeholder groups, representing advocacy groups as well as industry, also play a significant role in this process.

What emerges from this multi-institutional system is something other than a neat, unified approach to the basin, but one that increasingly appears to embody some core principles of sustainability. It suggests that sustainable development is more than an ethereal construct and can both take institutional form and begin to foster environmental improvement. This appears evident even in a region as physically vast, politically diverse, and institutionally fragmented as the Great Lakes Basin. But it further indicates that conventional depictions, reliant upon either a potent Leviathan or some locally based groundswell of creativity and collaboration, may exaggerate and oversimplify an exceedingly complex, multi-faceted phenomenon.

At the same time, this analysis will emphasize that any move toward sustainability within the basin is very much a work in progress. Serious impediments serve both to limit the impact of current initiatives and to endanger their long-term viability. In particular, the multi-institutional system that has evolved remains vulnerable to uneven—and ever-changing—levels of commitment from its respective jurisdictions. As certain institutions contemplate a reduced commitment, it remains uncertain whether other institutions have the inclination or capacity to fill the void. Indeed, the links between numerous policy initiatives remain very tenuous, often dependent on policy entrepreneurs who cannot be relied on to serve as permanent champions. Moreover, as we shall see, even the measurement of environmental quality in the basin continues to prove elusive, with no standard metric yet in place to serve as a reliable and comprehensive evaluative tool. These enduring impediments will be the primary focus of the concluding section of this chapter, offering cautionary reminders of both the complexity of the task at hand and the difficulty of sustaining recent progress.

Steps toward Sustainability

Many of the key concepts now associated with sustainability in the Great Lakes Basin were not minted freshly in the 1987 Brundtland Commission report. Ideas closely related to pollution prevention and regulatory integration across medium boundaries were clearly evident in the creation of expanded Canadian and American environmental policy systems in the basin during the late 1960s and early 1970s. These ideas tended, however, to be very general in nature, difficult to translate into policy. As a result, they were downplayed in both nations in favor of medium-based, command-and-control systems during these years, consistent with the general North American experience in the first environmental epoch. But if not entirely novel two decades or more later, these ideas were both refined over time and proved persuasive to a growing number of individuals with influence on the shape of environmental policy within the basin. This evolution took increasing institutional form as basin policy making entered the third environmental epoch.

The growing interest in and legitimacy of alternative approaches within the basin proved increasingly appealing to multiple levels of government and multiple institutions within the basin. There was no singular disaster or "focusing event" that triggered public and policy maker attention and directly prompted a shift in policy focus. Episodes such as the Lake Erie collapse in the late 1960s, widely oscillating lake levels throughout the 1980s, and the growing concern over toxics contamination in the late 1980s and into the 1990s all contributed to a sense of serious problems in the lakes and fostered a climate receptive to new policy steps. But these lacked the "big-bang" focus so common in policy formation models, whereby riveting events such as the Three Mile Island, Love Canal, and Exxon Valdez debacles clearly trigger substantial public alarm and foster new policy initiatives (Baumgartner and Jones 1993; Birkland 1997). Instead, a series of problems steadily served notice that the basin warranted serious attention and that extant approaches suffered from demonstrable deficiencies.

This growing recognition of problem intensity and complexity led multiple institutions to make periodic responses. Many of these were mutually reinforcing, even if not formally or systematically linked. At the same time that a binational authority would endorse cross-media integration and propose methods for pollution prevention, both federal governments and various states and provinces began to look seriously at programmatic options for promoting prevention and integrating environmental regulatory programs. Stakeholders representing industry and advocacy groups rarely assumed identical stances on key policy questions, yet moved many of their positions in remarkably similar directions over time. In many of these institutional settings, environmental policy professionals demonstrated growing conversance with the complexity of issues involved, a common recognition of the inherent limitations of prevailing approaches, and a willingness to explore and endorse alternatives. In many of these settings, such professionals demonstrated a capacity to take risks, learn across disciplinary, medium, agency, and intergovernmental lines, and begin to give some semblance of structure to what had once appeared to be little other than a set of slogans (Schneider and Teske, with Mintrom 1995; Kingdon 1984). By the mid-1990s, a network of such professionals constituted an informal community of experts with fairly common understandings of the complexity of existing environmental problems, and of policy alternatives with integrative and preventive emphases. In concert, different policy professionals sharing views on the future direction of environmental policy in the basin began to influence a policy shift aimed toward sustainable development goals from their respective institutional bases (Rabe 1996).

The Binational Role

None of the regional or binational entities with some responsibility for environmental policy in the Great Lakes Basin are dominant forces in guiding policy in a more sustainable direction. They generally lack regulatory powers to coerce or resource bases to entice broad-scale cooperation. Nonetheless, regional policy pronouncements such as a series of Great Lakes Water Quality Agreements (GLWQAs) and prominent reports from binational institutions such as the International Joint

Commission have consistently served to elevate public awareness of the complexity of regional environmental challenges. Moreover, they have provided early and visible endorsement of far-reaching policy reforms consistent with the tenets of the third environmental epoch.

The first GLWQA, signed by U.S. and Canadian government representatives in 1972, was firmly lodged in the tradition of the first environmental epoch. It focused largely on major point sources of water pollution, from industry and municipalities, in parallel with new national initiatives in the United States and, to a lesser extent, Canada. The 1972 Agreement was also responsive to the problem of nutrients that triggered excessive growth of algae and similar plants, which at the time was the predominant water quality concern in several of the lakes. Six years after this agreement, a revised GLWQA reflected a much expanded scope of concern, and a philosophy consistent with the third environmental epoch. This new agreement addressed the more complex problem of toxic contamination in the basin and endorsed the proposal that "virtual elimination" of such toxics should be a central goal of future basinwide policy. At the time, this was a particularly bold departure from more conventional pollution control strategies prevalent in the basin and North America more generally. The 1978 Agreement also introduced officially the concept of an "ecosystem approach" for regional policy, emphasizing that Great Lakes policy decisions must "focus on the physical, chemical, and biological relationships among air, water, and land." This emphasis was reiterated in the 1987 Agreement and was linked with an expanded emphasis on the cross-media dimensions of such basin problems as groundwater pollution, contaminated sediments, airborne toxics, and nonpoint source pollution.

These agreements did not translate into specific legislative proposals imposed uniformly across states and provinces in the basin. But they did give unprecedented visibility to these integrative and preventive approaches and indicate significant movement toward sustainability goals in advance of other North American regions, which remained more focused on pollution control and regulatory efficiency through the 1980s. They also created an institutional mechanism for overseeing regional performance by designating the IJC as the institutional watchdog over all of the agreements. This charge served to transform the IJC from its historic status as a near-moribund entity largely responsible for monitoring the levels and flows of boundary waters into a prominent advocate for basin sustainability.

The IJC has developed considerable analytical capacity, issuing regular reports on environmental quality issues within the basin and evaluating U.S. and Canadian adherence to agreement goals. Moreover, it has conducted a series of biennial meetings that draw substantial public participation and maintain a concerted focus on regional environmental concerns. These meetings are then followed by release of biennial IJC reports that often target particular problems. Recent reports have served to bring new attention to the environmental health ramifications of toxics contamination in the basin. In 1992, the IJC report recommended that both nations "develop timetables to sunset the use of chlorine and chlorine-containing compounds as industrial feedstocks" (Schwartz 1994, 493). Two years later, the IJC bolstered its previous call for virtual elimination with an endorsement of a "reverse onus" approach to chemical use and release in the basin. Its proposal would require that a substance be proven to be both non-toxic and non-persistent before being approved for use.

Both U.S. and Canadian governments have been free to ignore such recommendations, and both have used such latitude with frequency. In fact, the more controversial nature of some IJC proposals may have served to undermine its bases of budgetary and technical support, which the respective federal governments grant. The persisting role of IJC as "official nagger" has, however, served periodically to breathe life into the GLWQAs and stimulate a series of other proposals within the basin. The very language employed in the agreements and IJC pronouncements regularly finds its way into other policy documents, including those at the state and provincial level, and has maintained a fairly high level of visibility for these concepts across the basin.

The IJC has also been given oversight authority for a series of institutional innovations developed to address key provisions of the 1987 agreement. The most important of these are Remedial Action Plans (RAPs), which were mandated to help remediate forty-three of the most heavily contaminated areas around the basin. The agreements call for diverse stakeholders to complete a careful review of environmental evidence from all media, and foster strategies and coalitions that can transcend the fragmented, medium-based approaches of prior regulatory efforts. Plans are to go through a three-stage process, with each step involving IJC review, leading to eventual "delisting" of each area once contamination has been eliminated.

RAP implementation has been highly uneven to date, due in part to difficulty in securing resources to fund essential functions (Hartig and Zarull 1992). However, some RAPs have taken significant steps, both developing creative remediation plans and pursuing plan implementation (MacKenzie 1996). One RAP, Collingwood Harbour in Ontario, has met most key goals and was delisted in 1994, upon being found to "no longer have the attributes of an Area of Concern" (AOC) (Environment Canada 1996, 8; Krantzberg and Houghton, 1996). The Collingwood RAP process led to successful removal of contaminated sediments, and reversal of eutrophication and losses of fish and wildlife habitat. The RAP also resulted in new pollution prevention and water conservation initiatives for the area (IJC 1995, 177). Several other RAPs continue to demonstrate similar progress and have entered latter stages of their AOC remediation processes.

The Federal Role

The creative steps toward Great Lakes Basin protection promoted by binational action have, in many respects, been complemented by the actions of federal institutions, particularly in the United States. The respective roles of the U.S. and Canadian federal governments in the Great Lakes Basin represent significant contrasts in environmental federalism (Harrison 1996). In the United States, several policy initiatives of the federal government have clearly contributed to environmental improvement in the basin. Launched at various stages of the past quarter-century, these steps have helped move basin policy in a more preventive, integrative direction. In contrast, the Canadian federal government imprint is much more difficult to discern in the Great Lakes, due in large part to the continuing Canadian commitment to devolve most environmental policy functions to individual provinces. Ottawa has not made basin efforts comparable to those of Washington, D.C., and, as I will discuss, neither Ontario nor Quebec have filled this gap.

These divergent federal cases suggest that federal governments may have extremely important roles to play in fostering regional coordination across jurisdictional lines. Indeed, for all the opprobrium commonly heaped on U.S. federal involvement in environmental policy, its actions in the Great Lakes Basin suggests that it has played a number of constructive roles. This is evident both in early efforts to address pressing basin problems that were not being handled effectively by states or localities and, more recently, in finding new methods to both promote more sustainable policy approaches and find common ground among individual states.

Water Quality Initiatives
The first significant U.S. federal intervention on behalf of the Great Lakes involved its medium-specific approach to major point sources of water pollution in the 1970s. A combination of tough regulatory standards on major industry and an ambitious effort to improve sanitary sewers paid huge dividends in the region. In many respects, these sorts of pollution sources were not new. Many histories of regional economic development point to huge human health problems stemming from such contamination. In 1885–1886 alone, more than 80,000 Chicagoans died from typhoid, cholera, and dysentery after drinking contaminated water. As late as 1971, only 5 percent of the basin population living in residences with sewers were served by adequate treatment. The U.S. federal government responded aggressively in the 1970s with new regulatory standards and, in the case of sanitary sewage, major grant programs that led to substantial improvements in related areas of water quality.

This step not only paid immediate dividends but also helped clear the way for later initiatives involving the federal and other levels of government. In contrast, the Canadian federal government has never taken comparable steps in this area, consistently deferring to the prerogatives of individual provinces. Only in more recent years have Canadian provinces begun to emulate the U.S. approach to this aspect of water quality improvement. Ontario enacted its Municipal-Industrial Strategy for

Abatement (MISA) in 1986, borrowing substantially from the federal program that was initiated more than a decade earlier in the United States. MISA implementation has been consistently delayed, in large part due to cost concerns, and has served to delay adoption of more sustainability-oriented initiatives in the province.

At the same time, Ontario was only beginning to tackle major point sources and sanitary sewage, the U.S. federal government was passing through epoch two, emphasizing regulatory efficiency, and moving toward the next set of environmental challenges. Individual states began to assume an expanded role in many areas of policy initiation and implementation, but the federal role evolved in several key directions. In fact, different types of federal strategies in the past decade have clearly contributed to new steps in pollution prevention and regulatory integration across both medium and jurisdictional boundaries. And new medium-specific programs, such as the 1990 Clean Air Act Amendments, are expected to produce significant reductions in many air toxic sources from within and outside the basin that would otherwise land in one of the lakes. Neither federal nor provincial air quality legislation in Canada begins to approach this new federal legislation in terms of likely reduction of toxic emissions.

One of the most significant U.S. federal government interventions focused on the Great Lakes in recent years has involved efforts to unify water quality regulations among the eight states and various Native American tribes located in the basin. This unification has placed particular emphasis on setting stringent standards for releases of toxic contaminants deemed to pose unusually serious ecological and environmental health risks. As in the case of the earlier federal water pollution control initiative, this latter effort, best-known as the Great Lakes Initiative (GLI), stems from the failure of individual states to reach common ground. Enormous variations in state standards and growing concern over toxic contamination in the lakes prompted the eight governors of Great Lakes states to sign the Great Lakes Toxic Substances Control Agreement (GLTSCA) in 1986. This agreement endorsed the concept of "virtual elimination" of the release of persistent toxic substances into the lakes, following the 1978 lead of the GLWQA. It also set forth a number of commitments to move toward a more unified approach to this problem, including an effort to eliminate state-by-state disparities. Some states with particularly stiff standards, such as Michigan, were especially active in promoting this accord. The premiers of Ontario and Quebec also became signatories to the agreement through a separate Memorandum of Understanding in 1988.

Progress toward the goals set forth in the GLTSCA was exceedingly sluggish, however, prompting regional representatives in Congress to secure passage of the 1990 Great Lakes Critical Programs Act. This legislation called upon the U.S. Environmental Protection Agency to publish a Great Lakes Water Quality Guidance that would specify uniform numerical limits for all designated toxic pollutants to be released into the basin. The water quality guidance was also expected to specify minimum water quality standards for the entire basin, as well as establish antidegradation policies and implementation procedures. Under the legislation, states were to adopt all provisions of the guidance within two years of its publication, leaving the EPA as implementing agent in the event of state failure to meet this commitment.

The EPA released its guidance, widely known as the Great Lakes Initiative (GLI), in March 1993. Its rigorous standards went beyond conventional reliance on bioconcentration factors, which account solely for direct uptake of toxic chemicals from water. Instead, the GLI relied on bioaccumulation factors, accounting for both chemical uptake from water, as well as from the food chain. This shift in emphasis stemmed from growing evidence of profound bioaccumulation in Great Lakes fish, which if consumed have been demonstrated to pose serious environmental health risks. It further reflected unprecedented efforts to incorporate human health, aquatic life, and wildlife considerations into the process of setting regulatory standards. These new standards were to be imposed on all 3,795 direct dischargers into the U.S. side of the basin, ranging from steel and auto plants to publicly owned treatment works.

The guidance imposed standards that affected every state in the basin, even those already having established their own rigorous standards. However, both the initial formulation of the guidance and its refined versions have reflected an ongoing dialogue with states, regulated parties, and environmental groups, consistent with a general shift toward more collaborative intergovernmental governance reflected in later environmental epochs (Weber 1998). This review process has resulted, in many instances, in significant modifications incorporated into the final version published in March 1995. As one EPA official active in the process noted: "There was a perception going into this process that we would all go to the least common denominator regulation. That's not the case. We took good things from all the states and all of them will have to change. We've wound up with a much better way of doing business" (Askari 1994).

Despite the EPA's willingness to alter some important aspects of the guidance, many state and industry leaders have been very critical, alleging that the GLI represents yet another example of federal regulatory excess. Ironically, several of the states most supportive of the federal legislation that led to the GLI have now joined the ranks of critics. In particular, Republican governors John Engler of Michigan and Tommy Thompson of Wisconsin have led a chorus of others in urging the EPA to back away from the GLI and defer to individual state prerogatives. In particular, GLI opponents contend that the tough new regulatory standards will serve to deter economic development in competition with neighboring states—and provinces such as Ontario and Quebec—that are not similarly bound. Much of this change of heart stems from the fact that many of the states in the basin have undergone an ideological shift in gubernatorial and legislative leadership in the past half-decade. These changes may reflect backtracking in some basin states in the late 1990s, from sustainability goals and a renewed focus on regulatory efficiency. Elected state leaders from both major parties who supported the GLI process and would likely have endorsed the results are, in many instances, no longer in office. Moreover, even those earlier state supporters of the 1986 GLTSCA and the GLI proved unable to move collectively beyond the stage of mere pronouncement. Nonetheless, the GLI was implemented in 1999, and is widely perceived as a future contributor to reduced toxic loads in the basin. It reflects concerted federal action on behalf of the basin once states proved ineffective at implementing the broad policy goals they proclaimed.

Measurements of Ecosystem Quality

The U.S. federal government has also made a significant contribution to the basin through generation of an information source that has proven instrumental in a host of pollution prevention and regulatory integration initiatives. An enduring problem in environmental policy, and in formulating efforts to promote sustainability, is measurement of ecosystem quality. Conventional data sources tend to be medium-specific, inconsistently gathered across jurisdictional boundaries, and accessible only to individuals with considerable technological prowess. They tend to lack any potential to serve as a common metric to evaluate basinwide trends or direct creative policy initiatives. In contrast, the Toxics Release Inventory (TRI) has emerged as the closest thing to such a metric yet created in North America. Enacted as a right-to-know provision of the 1986 Superfund reauthorization bill, this mandatory information disclosure program led to the publication of the first national inventory on releases in 1988. Expanded significantly in 1990 through the federal Pollution Prevention Act, the TRI results in publication of an annual report on toxic releases to all environmental media for all states on the U.S. side of the basin.

Early TRI inventories drew considerable public attention to the issue of toxic releases but were somewhat unreliable as many complying parties struggled to adjust to the new provisions. In many cases, firms had no incentive or awareness of such releases, given their tendency to focus on individual point sources to meet specific permit requirements. Many such firms, embarrassed by TRI disclosures, began to take pollution prevention alternatives more seriously than before. In more recent years, the TRI has been refined, and now it produces extremely useful information. All TRI data for the basin can easily be stored within a standard microcomputer and are designed to be accessible to the public. Firms generating designated amounts of 623 toxic chemicals (increased in 1996 from the original list of 327 chemicals) are required to complete annual disclosure reports. Due to the 1990 revisions, participating firms must also provide information on toxic chemical source reduction and recycling. This latter information gives a clear indication of whether reductions in releases are due to actual reductions consistent with pollution prevention goals or, instead, reflect paper compliance. The expanded body of TRI data has proven instrumental for individual basin states, such as Minnesota, in attempting to devise creative new approaches for pollution prevention planning.

The Canadian government followed suit with its own inventory in 1993. The National Pollution Release Inventory (NPRI) is modeled closely after the TRI, although it involves far fewer total chemicals, does not require comparable source reduction and recycling information, and retains greater provision for protecting firm confidentiality. Nonetheless, the NPRI, in concert with the TRI, has begun to provide the first basin-wide picture of toxic chemical contamination. The first combined U.S.-Canadian report, published in 1995, revealed that nearly 700 million pounds of specified chemicals were released in the basin in 1993 (Environment Canada 1995). About 60 percent of those materials were released directly to the air, land, and water, with the remainder either landfilled, incinerated, recycled, or sent to municipal sewage plants for treatment. The initial report revealed unexpectedly high levels of release into Lake Erie and into the medium of air. Analysis of this data indicates that about 30 percent of the releases came from the Canadian side of the border and that Canadian releases were 5.94 kilograms per person versus a U.S. release rate of 4.96 kilograms per person (see table 8.2). State and provincial rates varied markedly, with Indiana, Ohio, and Ontario at the high end and Minnesota, Illinois, and Wisconsin at the low end. Differences in industrial mix and density, and numerous other factors, complicate straightforward interpretations of this data. However, Ontario releases are more likely to be significantly underestimated than those of the states in this comparison, due to far fewer chemicals listed for reporting, much lower rates of facility participation in the program, and lag in reporting submission dates that understates U.S. release reductions in the period under consideration (Rabe 1997).

The TRI has also served as the metric for a novel challenge regulation program launched by the U.S. federal government in 1989. The so-called 33/50 Program reflected a challenge by then-EPA Administrator William Reilly to firms with large annual TRI volumes to find ways voluntarily to reduce those levels. Firms accepting the challenge were expected to reduce releases for designated chemicals by one-third between 1988 and

Table 8.2
Toxic releases* in the Great Lakes Basin: 1992–1993 TRI and NPRI comparisons among basin states and Ontario

Jurisdiction	Percent of total basin releases	Percent of total basin population**	Releases of pounds per capita
Ontario	29.1	25.4	13.10
Michigan	21.9	27.9	8.99
Ohio	17.4	12.1	16.51
Indiana	15.5	4.3	41.45
New York	9.5	10.4	10.45
Wisconsin	4.8	7.6	7.21
Illinois	0.9	10.5	0.99
Pennsylvania	0.6	0.7	9.37
Minnesota	0.4	0.6	7.16
United States	70.9	74.6	10.93
Canada	29.1	25.4	13.10

* Releases include all on-site discharges of designated toxic chemicals into the environment. This includes emissions to air, discharges to water bodies, releases to land, and contained disposal into underground injection wells. The U.S. data are based on 1992 TRI reporting of 316 chemicals, and Canada data are based on 1993 NPRI reporting of 178 chemicals.
** Includes only areas of the state considered part of the basin, such as the northeast corner of Minnesota and the northwest corner of Pennsylvania.
Sources: Environment Canada and United States Environmental Protection Agency.

1992, and by one-half by 1995. The challenge focused in particular on seventeen TRI chemicals known to be highly toxic, released in large quantities, and thought to be highly promising for pollution prevention interventions. There is considerable debate as to whether this program has led to meaningful reductions, or largely just produced cosmetic changes in industrial practice (Arora and Cason 1995; Press and Mazmanian 1997; O'Toole, et al. 1997). However, the program appears to have had notable impact in the Great Lakes Basin, with total reductions of 188 million pounds of toxic chemicals from participating firms in the first four years alone (IJC 1995). Reductions appear to have been most significant in states with active pollution prevention initiatives, such as

Minnesota and New York. As an IJC analysis of the 33/50 Program noted, "Not all of these reductions are due to activities within the bounds of pollution prevention; however, they do represent a very significant achievement" (IJC 1995). The 33/50 Program has triggered a series of state- and industry-specific challenge initiatives within the basin. Most of these have utilized TRI and NPRI data as the common currency of evaluation, and many indicate significant promise.

These distinct U.S. federal government initiatives have been supplemented with additional steps. The EPA created a Great Lakes National Program Office (GLNPO) in the late 1970s, primarily to oversee U.S. compliance with the GLWQA. Over time, GLNPO has expanded its duties, due in part to the 1990 Great Lakes Critical Programs Act and the 1990 Pollution Prevention Act. It has played a central role in the development of the GLI, serving as an intermediary between states, specific interests within the basin, and both regional and central offices of the EPA. Since the eight states in the basin are divided into three separate EPA regions (New York is in Region 2, Pennsylvania is in Region 3, and Illinois, Indiana, Michigan, Minnesota, Ohio, and Wisconsin are in Region 5), GLNPO's coordination role has been extremely important, assuring that there is a basinwide advocate within a highly fragmented agency. It has also worked closely with individual states and industries on various pollution prevention initiatives and continued implementation of more recent GLWQAs. GLNPO and other EPA offices have also distributed substantial funding to individual states for pollution prevention and related regulatory integration activities. Great Lakes states are consistent with those from other regions in relying on federal grant dollars for nearly half of their total pollution prevention expenditures, although there is considerable variation among individual basin states in this regard (GAO 1994; Barnes 1994; Doyle 1995).

Interstate Institutions

Neither the U.S. nor Canadian constitutions were drafted with serious consideration of mechanisms to facilitate problem solving among neighboring states or provinces. Water basins pose a particularly complex question of interjurisdictional governance, as they tend to render political boundaries meaningless. One environmental historian has noted, "Many of the environmental crises of the 20th Century have resulted from externalities generated by artificial political boundaries" (Thorson 1994, 140). U.S. states have begun to respond to this problem through use of interstate agreements, establishing on at least thirty-five occasions formal compacts for joint oversight of common water management. There is no comparable formal process yet to emerge among provinces in Canada. However, there is a growing number of agreements that involve two or more provinces over a particular environmental or natural resource concern, and provinces have on occasion signed onto U.S. state agreements.

Interstate compacts and region-specific state organizations have provided mechanisms to foster regional coordination and, in some respects, complement comparable efforts launched by binational and national authorities. Basin water quantity and the concern over the fluctuating levels of individual lakes have continually triggered interstate interest in devising new institutions to promote research and facilitate consensus building. As early as the 1950s, these concerns led to the development of the Great Lakes Commission as a formal advisory and advocacy body. In turn, the Council of Great Lakes Governors was created in the early 1980s, to provide a regular forum for basin governors to explore cooperative approaches to economic development and environmental protection.

These institutions lack the authority of federal, state, and provincial governments and therefore have relied on a series of largely nonbinding agreements, often called charters. For example, the Council of Great Lakes Governors oversaw multistate negotiations, also involving Ontario and Quebec, that led to ratification of a Great Lakes Charter in 1985. All signatories pledged "not to increase diversions of consumptive uses of basin water without first seeking the consent of all states and provinces" (Thorson 1994, 161). This step was triggered in large part by an external threat in the late 1970s: a proposal to transfer substantial quantities of basin water for use in a national coal-slurry pipeline. The pipeline proposal was eventually abandoned, but the threat of potential diversions endures. Some states continue to contend that certain neighbors, most commonly Illinois, repeatedly violate the charter with excessive diversions. The charter lacks any mechanism to monitor and enforce compliance, making it difficult to evaluate the merit of such claims. In this regard, it is similar to other types of agreements established by the Council of Great Lakes Governors. One such example is the Great Lakes Toxic Substances Control Agreement of 1986, which resulted in a broad endorsement of aggressive toxics reduction in the basin, but lacked any compliance mechanisms and ultimately was eclipsed by the EPA's Great Lakes Initiative.

A more far-reaching interstate pact has been undertaken during the 1990s, under the auspices of the Great Lakes Commission. Supported by the Chicago-based Joyce Foundation, the commission sought diverse, basinwide input into the drafting of an Ecosystem Charter for the Great Lakes-St. Lawrence Basin. The Charter was released in 1994, and continues to collect hundreds of signatories representing state, local, and federal agencies, environmental advocacy groups, industry associations, universities, and other organizations. It sets forth seventeen principles that fully embrace the concepts of sustainable development and ecosystem management, and offers guiding principles for all future economic and environmental policy within the basin. Specific provisions endorse virtual elimination of toxic substances and promotion of biological diversity as cornerstones of future policy. Other sections sketch broad goals, such as Principle 9, which states that "Societal needs for a healthy ecosystem and economy shall be addressed by promoting the sustainable use of renewable natural resources" (Ecosystem Charter 1994).

The Ecosystem Charter underscores both the promise and limitations of interstate action, as presently practiced. In broad language, it sets forth an ambitious course of action entirely consistent with the goals of sustainability articulated in the WCED report. But like other multistate efforts, such as the Great Lakes Charter, it relies on the goodwill of all parties to pursue its broad goals, lacking either a specific blueprint for action or political authority to secure collective action. Instead, it tends to leave compliance questions in the hands of individual states.

Single-State Institutions

No two states or provinces in the basin have responded in identical fashion to the charge set forth in the Ecosystem Charter. Reflecting their differing political cultures and levels of support for environmental protection, states and provinces vary enormously in both their degree of commitment and demonstrated capacity to pursue goals consistent with sustainability and ecosystem management principles. Nonetheless, every state or province in the basin has made some formal effort in this regard, most commonly through pollution prevention programs that offer technical assistance to participating industries. Some states have gone well beyond this point, experimenting with a wide range of initiatives that attempt to translate sustainability principles into institutional forms.

Minnesota emerges as the clear leader among basin states and provinces in this regard, having devised a host of pollution prevention and regulatory integration programs during the past decade. Minnesota has already served as a model for study and emulation by other states and provinces, and is unique in elevating these alternatives from experimental status to a guiding position in state environmental protection and economic development efforts. The Minnesota approach reflects a blend of incentives and regulatory measures, all linked by the common goal of minimizing environmental contamination, and explicitly directed at a statewide commitment to sustainability.

Building on earlier efforts to provide technical assistance to industry for pollution prevention, Minnesota enacted a far-reaching Toxic Pollution Prevention Act in 1990. This legislation stemmed from hearings that systematically reviewed existing barriers to pollution prevention and regulatory integration and attempted to find ways to eliminate them. The act established "pollution prevention at the source" as the guiding principle of all state environmental protection efforts and made clear that this emphasis would be translated into specific policy initiatives.

One cornerstone of the 1990 act was mandatory pollution prevention planning. Minnesota joined a small number of states, such as Massachusetts and New Jersey, in utilizing TRI data for facility-based planning programs. The state moved quickly to implement TRI provisions in the late 1980s, to maximize the likelihood that the data would emerge in reliable form. Expansion of TRI reporting in 1990 requirements provided additional information, which Minnesota required firms to incorporate into their planning processes. In turn, the state sets fees based on volume and type of toxic substances released, with resulting revenues providing both a disincentive to releases and a substantial source of funding for expanded technical assistance efforts. Facility plans are to be revised through annual progress reports, contributing to a systematic evaluation of progress in pollution prevention that remains novel among states and provinces in the basin.

Minnesota has supplemented its pollution prevention initiatives by revisiting virtually every step of the conventional regulatory process during the 1990s. Officials discovered considerable potential for inadvertent cross-media and cross-jurisdictional pollutant transfer through the implementation of conventional approaches. They have pursued more integrative approaches to monitoring, inspection, and enforcement through pilot projects concentrating on firms located along the shore of Lake Superior and in the Minnesota River Basin. Permitting has begun to be modified through creation of so-called flex permits, which offer participating firms substantial flexibility in determining how to reduce emissions in exchange for firm commitment to dramatic emission reductions and installation of sophisticated systems of continuous emissions monitoring (Rabe 1995). State officials have also worked creatively with federal counterparts in these initiatives, seeking ways to modify existing regulatory programs to promote approaches more consistent with the 1990 act. They have also been very effective in seizing on federal opportunities to explore such alternatives. For example, they have aggressively pursued federal pollution prevention grants and have become involved in the EPA's Project XL, which seeks ways to integrate pollution prevention and regulatory integration with facility-specific exigencies. Minnesota also established its own version of the federal 33/50 Program; the so-called Minn 50 Program resulted in extensive participation and enormous drops in toxic releases.

Virtually every available source of evaluation suggests that these efforts have contributed to significant improvements in environmental quality in Minnesota (Minnesota Pollution Control Agency 1996). The state has consistently registered some of the biggest drops in TRI reports among basin states, and has developed sufficient evaluative capacity to be able to pinpoint continuing problem areas and avenues of opportunity. Volumes of hazardous and solid waste generation have dropped more rapidly than in any other state or province in the basin. In response, the state continually attempts to build on its prior experience, increasingly turning its attention to nonpoint sources of contamination and generators of relatively small amounts of toxic contaminants. These numerous efforts are increasingly being integrated through statewide planning for sustainability involving an unusually diverse coalition of stakeholders.

Far more than a single-program experiment, Minnesota appears to have made huge strides in moving its environmental policy efforts in directions eminently compatible with sustainability principles. Among other states and provinces in the basin, New York and Wisconsin come closest to the Minnesota standard. Both states have extended histories of attempting to address cross-media concerns in devising regulatory approaches. New York has continued to build on this experience in more recent years with new waste reduction planning initiatives and efforts to integrate core regulatory tools. In turn, Wisconsin has experimented with a series of organizational innovations intended to foster greater integration across traditional boundary lines within the state Department of Natural Resources. Wisconsin has also attempted to develop integrative programs designed to address all likely contamination sources of specified watersheds (Adler, Landman, and Cameron 1993, 186). Other states, such as Ohio and Indiana, give less indication of following this line of innovation and, in turn, remain those units generating the highest toxic release rates per capita (see table 8.2). Michigan, Illinois, and Pennsylvania demonstrated considerable early promise in integration and prevention, well ahead of most neighbors in the 1970s and 1980s, but appear in recent years increasingly committed to regulatory efficiency reforms most closely associated with the second environmental epoch.

Enduring Impediments

Developments at binational, national, multistate, and state levels all indicate progress toward Great Lakes Basin sustainability. Institutions at each level have demonstrated some capacity to not only articulate a vision of a more comprehensive commitment to the region but also begin to translate those statements into specific policy initiatives. In some instances, one can begin to discern the actual components of a more integrative system, representing a big step from the exceedingly abstract discussions of this topic a decade or more ago. Indeed, many of the key tenets of the Brundtland Commission report appear to be reflected in actual shifts in Great Lakes Basin policy.

Basin policy remains, however, in a period of transition and uncertainty rather than in any definitive shift toward sustainability. Numerous conditions continue to impede the further development of cross-jurisdictional cooperation toward common environmental goals and, in fact, may serve to retard recent gains. Just as each institutional level has played some role in moving policy in the directions already noted, each also retains limitations that brake further progress. Recent scholarly analysis of regional approaches may serve to overstate the ease with which common commitment to resource protection can occur. Extensive scholarly work on the protection of "common-pool resources" sets forth generic design characteristics that are expected to lead multiple stakeholders to work cooperatively toward the common environmental good (Ostrom 1990; Ostrom, Gardner, and Walker 1994). However, the vast majority of examples cited in such accounts involve regions contained within a single nation and possessing total populations that constitute less than 1 percent of the total basin population. Moreover, potential externalities tend to be extremely limited in these cases.

In contrast, the basin constitutes a massive "community," involving enough people to exceed the populations of all but eleven nations, diverse jurisdictional and institutional actors, and externality problems galore. Its environmental protection remains a daunting collective-action challenge, even for governing institutions with substantial resources and creative energies. Its recent strides toward sustainability, given those realities, represent a remarkable success story. But as one review of basin progress concluded: "The good news is that the Great Lakes region is a leader in transnational comprehensive environmental management. The bad news is that this may say more about inadequacies elsewhere than about the health of the basin ecosystem" (Colborn et al. 1990, 229). Such an assessment tends toward overstatement, but does serve as a reminder of substantial impediments that endure. The following points illustrate significant stumbling blocks to further realization of the goals set forth in the Brundtland Commission report.

Limits to Central Guidance

Binational institutions such as the IJC have undergone a remarkable evolution in recent decades, becoming outspoken advocates for basinwide policy shifts compatible with sustainability goals. But their role remains largely symbolic, confined to exhorting national and subnational governments to respond to their pronouncements. They generally act only in response to specific requests from U.S. or Canadian federal governments, and often see their recommendations go ignored or only superficially implemented. Even policy innovations under IJC auspices, such as Remedial Action Plans (RAPs) and lakewide management plans, are almost completely dependent on the resources and cooperative goodwill of federal, state, provincial, and local governments. As a result, for every Collingwood Harbour-type breakthrough, more than a dozen other RAPs remain mired in the very early stages of evaluation.

Great Lakes Water Quality Agreements signed by U.S. and Canadian federal representatives further illustrate this phenomenon. GLWQAs appear, on paper, to represent far-reaching steps toward a unified approach to basin protection. Studies of national compliance with key GLWQA provisions show a consistent pattern of evasion and delay (Great Lakes United 1991; Royal Commission on the Future of the Toronto Waterfront 1993). Furthermore, survey research suggests that these types of policy guidelines may not even be examined, much less implemented, by key stakeholders in the basin (Marans et al. 1988). As one representative of a basinwide environmental group explained: "The problem is that the people who actually write the permits and implement the pollution control programs not only don't go to [IJC] meetings, but usually have no idea what the Great Lakes Water Quality Agreement is; they've never read it; they wouldn't know how to find it" (Rabe and Zimmerman 1992, 104–105).

Institutions and agreements such as the IJC and GLWQA contribute to sustainability goals through their jawboning techniques. As political scientist William Gormley has demonstrated, such approaches may serve to "promote bureaucratic creativity, flexibility, innovation, and experimentation. In effect, they emphasize long-term goal attainment rather than short-run compliance with hastily crafted directives" (Gormley

1989, 231). In the Great Lakes Basin, such efforts have provided high-level endorsement of concepts such as "virtual elimination" of toxics and sustainability, encouraging and prodding national and subnational governments to respond. Some of the policy shifts noted above are at least partially attributable to this binational input. But, as the uneven record of national, state, and provincial governments indicates, it is also eminently possible to ignore jawboning efforts.

Federal Government Shirking

Federal leaders in both Ottawa and Washington, D.C., have habituated the practice of offering immediate and energetic embraces of IJC or GLWQA policy pronouncements. However, these endorsements do not necessarily translate into shifts in federal policy. This lag is particularly evident in Canada, where the federal government not only maintains a minimalist presence in environmental policy but also remains overwhelmingly committed to medium-based, pollution control strategies. Canadian federal environmental policy continues to be remarkably dependent on prior U.S. federal experience. Consequently, Canada is just beginning to consider a number of promising alternatives. Its existing pollution prevention programs and mandatory toxic release disclosure process remain in very early stages and do not approach U.S. predecessors in breadth.

In instances in which Canada has apparently embraced more preventive and integrative approaches, these often prove very superficial. The launching of the so-called Green Plan in the early 1990s was heralded as a major breakthrough whereby diverse grant programs would be unified behind the common cause of sustainability. Subsequent analyses of the plan reveal a thinly disguised strategy more closely resembling conventional developmental policy initiatives, with little evidence of any significant shift in funding from traditional priorities (Hoberg and Harrison 1994). In turn, periodic efforts to modify major pieces of Canadian federal legislation, such as the Canadian Environmental Protection Act (CEPA), represent little other than cosmetic efforts to move beyond traditional approaches. A series of recent analyses concur that CEPA papers over historic divides rather than offers any bold new alternatives. "CEPA does not currently accommodate effective pollution prevention strategies and must be substantially amended in order to do so," noted a 1995 Environment Canada study (Environmental Protection Service 1995, 1). Similarly, a 1994 House of Commons study concluded that while, in theory, the legislation should provide "a key legislative base for promoting pollution prevention" and regulatory integration, in reality "a major change of emphasis would be required for CEPA to provide such a base" (Standing Committee on Environment and Sustainable Development 1995, xxiii). In virtually every area where U.S. federal and state governments have begun to devise policy alternatives more compatible with sustainability goals, Ottawa clearly lags behind.

The federal government contribution is somewhat more heartening on the U.S. side of the basin. Earlier and more aggressive water and air pollution control efforts enacted by the U.S. federal government led to significant environmental improvements and also paved the way for more preventive, integrative initiatives in recent years. As discussed earlier, the Great Lakes Initiative, Toxics Release Inventory, and a series of pollution prevention initiatives offer considerable potential for reaching sustainability goals. However, many key components of federal environmental policy in the United States continue to adhere to conventional, medium-based emphases. Enabling legislation, the organizational design of the EPA, and an agency culture that continues to foster turf protection all augur against more far-reaching evolution of federal policy. Moreover, both Canadian and U.S. federal environmental agencies face uncertain fiscal futures. Recent U.S. budget cuts have had a substantial impact on several basin-specific programs and have also threatened the implementation of the TRI. In turn, Environment Canada and related agencies face proposed cuts of approximately 30 percent over a three-year period, with those cuts likely to fall most heavily in newer areas such as those devoted to pollution prevention. Any future commitment of federal governments to areas with direct bearing on sustainability thus remains highly uncertain.

State and Provincial Government Shirking

In the absence of clear marching orders from binational or federal authorities, states and provinces bear the brunt of responsibility for securing collaborative approaches that best represent the region. As we have seen, most units in the basin have made some effort to move toward more preventive and integrative policies in recent years. Some, such as Minnesota, have unilaterally pursued a host of promising new policy departures. On the whole, however, individual states and provinces continue to face considerable incentives to shirk responsibility for basinwide well-being. Subnational units working to reduce emissions and waste volumes may largely be serving neighboring states and provinces, given prevailing currents of pollutant transfer and patterns of waste export. They may, in short, have limited incentive to act on their own, especially when their own environmental problems may endure due to contamination transferred from other jurisdictions. Given the economic and political costs associated with imposing new requirements on regulated parties, it may be enticing for individual states or provinces to stick with traditional routines. State and provincial caution in these matters may also be reflected by some late-1990s efforts to cut environmental staff and total expenditures significantly and propose "customer-friendly" regulatory alternatives that may constitute a new form of efficiency emphasis so central to the second epoch. Such a shift is most visible in Ontario and Michigan, each of which has reduced environmental agency staff by more than 20 percent since 1996, and has focused almost exclusively on easing regulatory burdens.

Recent experience in the basin suggests that while progress has been made on some fronts, the search for unified, basinwide strategies has been difficult for states and provinces. There is little sense of the convergence of perceived interests that is so central to common-pool resource cases, except in those instances in which a particular problem is widely perceived as a common threat, such as in blocking water diversions out of the region. Substantial evidence suggests tremendous variation in individual state and provincial standards, which is reflected in the necessity of federal intervention through the GLI to eliminate persistent variations in state water quality standards. Until the GLI is implemented, the basin's lakes will continue to be bordered by states imposing very different standards. For example, New York allows each permittee to discharge in excess of 145 kilograms per year of mercury while Michigan would restrict the same regulated party to less than 1 kilogram per year. Similarly, Michigan would limit permittee discharge of lead to 700 kilograms per year, whereas Illinois's allowance for the same discharger would reach 7,000 kilograms per year (Adler, Landman, and Cameron 1993, 163).

Beyond formal standards, there also appears to be enormous variations in the rigor and accuracy of state and provincial monitoring and enforcement activities in the basin. In turn, some states, most notably Minnesota, conduct systematic reviews of all programs, including those related most directly to prevention and sustainability goals. Other states and provinces largely refrain from evaluation other than to herald individual success stories (GAO 1994, 46–47; Doyle 1995). The use and accessibility of TRI data provides a telling illustration of this variation. Whereas some states have seized upon TRI as a valuable tool in devising new policies and facilitating improved evaluation of existing ones, others have viewed it as a federally mandated nuisance and do little other than meet minimum data-gathering requirements. It is too soon to tell which pattern Ontario and other provinces will follow as NPRI data become available, although their recent records suggest little likelihood that they will actively use such data in pursuit of more preventive and integrative alternatives. Massive cutbacks in Ontario funding of environmental programs during 1997 may also limit further provincial development of NPRI and related policy changes, reflecting a growing emphasis on regulatory streamlining rather than sustainability. Enormous variation, of course, is also found in per capita rates of toxic release by jurisdiction, and many of the states with only tiny percentages of their territory within the basin have the lowest per capita release rates (see table 8.2).

Interstate variation is also reflected in the enduring problem of cross-boundary transfer of pollutants and wastes. In the United States, federal policy has periodically attempted to address serious pollution "balance-of-trade" problems after states failed to resolve such difficulties on their own. States like Illinois, Ohio, and Pennsylvania have continued to contribute substantial quantities of air pollution, much of which becomes land or water pollution problems elsewhere in the basin or in other regions. For example, nearly half of the sulfate levels found in New York's Adirondack Mountains emanates from Midwestern air pollution sources. As political scientist William Lowry has noted, the current federal structure "creates incentives for states to export pollution" (Lowry 1992, 44).

Management of hazardous and low-level radioactive wastes in the basin further illustrates this phenomenon, at times resembling a shell game, whereby individual states view out-of-state export as their preferred strategy. Rather than explore creative waste reduction or waste treatment options, basin states often take an "out-of-sight, out-of-mind" approach. The fifteen-year experiment with decentralization of siting and most waste management functions has fostered considerable animosity among individual states. In low-level radioactive waste, siting efforts in Illinois, Michigan, and Ohio have been textbook examples of siting gridlock, leaving export to South Carolina and on-site storage the only options for dozens of waste generators within the region. In this area, Canadian provinces have been somewhat more successful, which is particularly evident in Ontario's cooperative 1996 agreement for a long-term waste management strategy.

While it proves relatively easy for governors and premiers of Great Lakes Basin states and provinces to find common cause in making bold pronouncements on behalf of the regional ecosystem, it remains difficult in many cases to translate those declarations into meaningful collective action. The basin may simply be too massive an entity for individual states and provinces to focus systematically on their potential contribution to regional ecological protection. Relatively limited sharing of information and ideas occurs among officials across state and provincial boundaries, one of a series of continuing challenges for decentralized approaches to basin well-being (Rabe and Zimmerman 1992).

External Threats

Even if all individual governing units in the basin were successful in banding together in pursuit of sustainability, the environmental problems of the region would not simply disappear. Rather than a hermetically sealed entity, the basin remains vulnerable to a host of external threats that cannot be easily deflected. In recent years, two particularly serious threats have emerged, both of which defy easy measurement, much less remediation.

First, unwanted imports, most commonly in the form of air pollution, compound existing toxic contamination problems in the basin. Increasing evidence suggests that unexpectedly large percentages of select toxics that reside in the basin are not generated internally, but instead drift or float in from other regions. In some instances, all plausible sources of certain kinds of toxics are outside the basin. Certain pesticides and insecticides in use only in the U.S. South continue to accumulate in all five of the lakes. Even new quantities of DDT, long banned in both Canada and the United States, keep arriving into the basin, with the most likely route being air currents from Central America, where this toxic substance remains in use.

Second, nonindigenous fish and animal species continue to be introduced into the basin, often disrupting food chains and posing significant remediation challenges. The so-called exotic species problem is not new to the Great Lakes, as the sea lamprey entered Lake Erie in 1921, and ultimately spread to all five lakes. The lamprey has had devastating effects on trout populations and has triggered an ongoing campaign to curb its further proliferation. But as shipping and trade passing through the basin increased, the pace of new exotics being introduced has accelerated in recent years. Ballast water of ships is often thought to be the initial source of these invaders. In more recent years, the zebra mussel has wreaked havoc in several areas of the basin, and a series of new exotics, such as the ruffe, quagga mussels, and round goby pose additional challenges completely unanticipated. Each new exotic not only constitutes a new potential disruption to the ecology of the Basin but also tends to shift resources away from other potential basin concerns. For example, basin municipalities and industries spent $120 million on largely ineffective zebra mussel control efforts between 1989 and 1994. Problems stemming from exotics also include an increasing number of nonindigenous plants, such as purple loosestrife and watermilfoil.

Conclusions

To paraphrase Mark Twain, reports anticipating the death of the Great Lakes issued a quarter-century ago have proven inaccurate. A number of important indicators of lake water quality and basin ecosystem health have stabilized or improved, despite increases in population, industrial activity, and automobile usage. Many of these salutary developments can be traced directly to interventions from various levels of government. These developments should not serve, of course, to obscure enduring challenges facing the basin. Coordination of efforts across governmental levels remains difficult, and individual states and provinces have very mixed records in addressing regionwide concerns. Many of the most promising new innovations remain in early stages of implementation, with their long-term performance or capacity for replication in other basin jurisdictions unclear. Moreover, the recent history of the basin reveals an ever-expanding set of challenges to ecosystem health. At the same time that external contaminants and exotic species continue to confound basin policy makers, there is growing indication that the next round of internal policy development must focus on ubiquitous sources of contamination from within the basin. Far beyond the traditional emphasis on large point sources of pollution, such as major industry and municipalities, nonpoint water pollution may be the next—and most substantial—challenge yet faced. Nonpoint sources include virtually all residents in the basin, including farmers and home owners, whose penchant for land development and pesticides invariably compound water quality problems.

Viewing environmental sustainability in the basin from this perspective, and attempting to shift the focus of its citizenry toward sustainability goals, remains a formidable challenge. Nonetheless, the past three decades of experience within the basin reflects an evolution largely consistent with the epochs outlined in chapter 1. After earlier emphasis on command-and-control strategies and experimentation with regulatory efficiency measures, many areas of basin policy are making a clear shift into the third epoch, and reflect its emphasis on sustainable development. Most notably, one can discern a movement from pollution control strategies toward more preventive and integrative efforts in recent years. Such contributions toward this transformation are evident at binational, federal, and state and provincial levels of governance within the basin. In some instances, new systems are now being tested that appear to hold considerable promise for placing environmental policy efforts into a more comprehensive context and for confronting some of the most serious challenges to basin sustainability. This transition is not seamless, and could indeed be undermined by a variety of factors, such as possible reluctance of individual states and provinces to participate fully in this collective effort and a reversion to regulatory efficiency goals so central to the second epoch. Nonetheless, the Great Lakes Basin, despite its physical breadth, population density, and industrial intensity, demonstrates the potential on a regional basis for supplanting conventional approaches to environmental protection with an emphasis on defining and pursuing sustainability.

References

Adler, Robert W., Jessica C. Landman, and Diane M. Cameron. 1993. *The Clean Water Act 20 Years Later.* Washington, DC: Island Press.

Arora, Seema, and Timothy N. Cason. 1995. "A Voluntary Approach to Environmental Regulation: The 33/50 Program." *Resources* (Summer): 6–10.

Askari, Emilia. 1994. "Great Lakes Deal Has Long Reach." *Detroit Free Press* (December 25): 81.

Baumgartner, Frank, and Bryan Jones. 1993. *Agendas and Instability in American Politics.* Chicago: University of Chicago Press.

Birkland, Thomas. 1997. *After Disaster.* Washington, DC: Georgetown University Press.

Barnes, Laura L., ed. 1994. *The Pollution Prevention Yellow Pages.* Washington, DC: National Pollution Prevention Roundtable.

Caldwell, Lynton K., ed. 1988. *Perspectives in Ecosystem Management in the Great Lakes.* Albany: State University of New York Press.

Colborn, Theodora E., Alex Davidson, Sharon N. Green, R.A. (Tony) Hodge, C. Ian Jackson, and Richard A. Liroff. 1990. *Great Lakes, Great Legacy?* Ottawa: Institute for Research on Public Policy.

Doyle, Mary Beth. 1995. "State-Level Pollution Prevention Initiatives." Unpublished M.P.H. Thesis, University of Michigan School of Public Health.

Durnil, Gordon K. 1995. *The Making of a Conservative Environmentalist.* Bloomington: Indiana University Press.

Environment Canada. 1995. *Industrial Releases Within the Great Lakes Basin: An Evaluation of NPRI and TRI Data.* Ottawa: Environment Canada.

Environment Canada. 1996. *First Progress Report Under the 1994 Canada-Ontario Agreement Respecting the Great Lakes Basin Ecosystem.*

Environmental Protection Service, Environment Canada. 1995. *Environmental Protection Legislation Designed for the Future—a Renewed CEPA.* Ottawa: Environment Canada.

Gormley, William T., Jr. 1989. *Taming the Bureaucracy.* Princeton: Princeton University Press.

Great Lakes Commission. 1994. "Ecosystem Charter for the Great Lakes-St. Lawrence Basin." Ann Arbor: Great Lakes Commission.

Great Lakes United. 1991. *Broken Agreement: The Failure of the United States and Canada to Implement the Great Lakes Water Quality Agreement.* Buffalo: Great Lakes United.

Harrison, Kathryn. 1996. *Passing the Buck: Federalism and Canadian Environmental Policy.* Vancouver: University of British Columbia Press.

Hartig, John H., and Michael A. Zarull, eds. 1992. *Under RAPs: Towards Grassroots Ecological Democracy in the Great Lakes Basin.* Ann Arbor: University of Michigan Press.

Hoberg, George, and Kathryn Harrison. 1994. "It's Not Easy Being Green: The Politics of Canada's Green Plan." *Canadian Public Policy* 20, No. 2: 119–137.

International Joint Commission. 1995. *1993–95 Priorities and Progress Under the Great Lakes Water Quality Agreement.* Windsor: IJC.

John, DeWitt. 1994. *Civic Environmentalism.* Washington, DC: Congressional Quarterly.

Kingdon, John. 1984. *Agendas, Alternatives, and Public Policies.* Boston: Little, Brown and Company.

Krantzberg, Gail, and E. Houghton. 1996. "The Remedial Action Plan That Led to the Cleanup and Delisting of Collingwood Harbour as an Area of Concern," *Journal of Great Lakes Research* 22, 2: 469–483.

Lowry, William. 1992. *The Dimensions of Federalism.* Durham: Duke University Press.

Marans, Robert W., Jonathon W. Bulkley, Rujiroj Anambutr, Jinsheng Fan, and Susan MacKenzie. 1998. *Trends and Emerging Environmental Issues in the Great Lakes.* Ann Arbor: Michigan Sea Grant College Program.

Minnesota Pollution Control Agency. 1996. *Protecting Minnesota's Environment: A Progress Report.* St. Paul: Minnesota Pollution Control Agency.

Ophuls, William, and A. Stephen Boyan, Jr. 1992. *Ecology and the Politics of Scarcity Revisited: The Unraveling of the American Dream.* New York: W.H. Freeman.

Ostrom, Elinor. 1990. *Governing the Commons.* Cambridge: Cambridge University Press.

Ostrom, Elinor, Roy Gardner, and James Walker. 1994. *Rules, Games, and Common-Pool Resources.* Ann Arbor: University of Michigan Press.

O'Toole, Laurence J., Jr. Chiluk Yu, James Cooley, Gail Cowie, Susan Crow, Terry DeMeo, and Stephanie Herbert. 1997. "Reducing Toxic Chemical Releases and Transfers: Explaining Outcomes for a Voluntary Program," *Policy Studies Journal* 25, 1: 11–26.

Press, Daniel, and Daniel A. Mazmanian. 1997. "The Greening of Industry: Achievement and Potential." In *Environmental Policy in the 1990s: Reform or Reaction?* eds. Norman J. Vig and Michael E. Kraft. Washington, DC: CQ Press. Pp. 255–277.

Rabe, Barry G. 1996. "An Empirical Examination of Innovations in Integrated Environmental Management: The Case of the Great Lakes Basin." *Public Administration Review* 56 (July/August): 372–381.

Rabe, Barry G. 1997. "The Politics of Sustainable Development: Impediments to Pollution Prevention and Policy Integration in Canada." *Canadian Public Administration* 40 (Fall): 415–435.

Rabe, Barry G. 1995. "Integrating Environmental Regulation: Permitting Innovation at the State Level." *Journal of Policy Analysis and Management* 14, 3 (Winter): 467–472.

Rabe, Barry G., and Janet B. Zimmerman 1992. *Toward Environmental Regulatory Intergration in the Great Lakes Basin.* Ann Arbor: Resource for Public Health Policy.

Royal Commission on the Future of the Toronto Waterfront. 1993. "Regeneration: Toronto's Waterfront and the Sustainable City."

Schneider, Mark, and Paul Teske, with Michael Mintrom. 1995. *Public Entrepreneurs: Agents for Change in American Government.* Princeton: Princeton University Press.

Schwartz, Alan M. 1994. "Canada-U.S. Environmental Relations: A Look at the 1990s." *American Review of Canadian Studies* 24 (Winter): 489–508.

Standing Committee on Environment and Sustainable Development, House of Commons. 1995. *Reviewing CEPA: The Issues,* No. 7. Ottawa: House of Commons.

Thorson, John E. 1994. *River of Promise, River of Peril.* Lawrence: University Press of Kansas.

U.S. General Accounting Office. 1994. "Pollution Prevention: EPA Should Reexamine the Objectives and Sustainability of State Programs." Washington, DC: GAO.

Weber, Edward. 1998. *Pluralism by the Rules: Conflict and Cooperation in Environmental Regulation.* Washington, DC: Georgetown University Press.

IV

Overview and Implications for a Sustainable Future

9

Conclusions: Toward Sustainable Communities

Michael E. Kraft and Daniel A. Mazmanian

As the twentieth century draws to a close, environmental policy in the United States is in the early stages of what could be a profound transition toward sustainable communities. It is too early to speak confidently of how far this transition will go, how fast it will occur, or how successful it will be in addressing the major environmental challenges facing the nation and the world. Yet there is no question that a shift has begun, from the era of environmental regulation that dominated the 1970s and 1980s, to a more diversified set of environmental policies rooted in the long-term goal of sustainability. At this historic juncture, choices about policy design, organizational change, democratic participation, and other issues in environmental governance depend critically on our understanding of the transformations now under way and their implications—for politics, the economy, society, and the environment. We hope this book contributes to that understanding and that it stimulates others to inquire about additional cases and circumstances that can fill in the many gaps in our knowledge about the transition to sustainability-based environmental policy.

In this concluding chapter, we consider the implications of the eight individual chapters for both theory and practice, assess emerging trends in environmental policy related to the arguments in these chapters, and suggest some fruitful lines of additional research that may help to confirm the findings in the case studies, or suggest how they should be modified. We also revisit our thesis about the nature of the transition toward sustainable communities and offer some observations about lessons learned, particularly the factors that seem most significantly to affect the character and pace of the transition to environmental sustainability.

Chapter Findings and Implications

The previous eight chapters have examined a variety of environmental policy developments within different regions, states, and communities. Six of these chapters consider the most pertinent issues raised by their case study while also speaking to the larger purposes of the book. At this juncture we need to consider the separate findings of these chapters in relation to one another and with respect to the framework introduced in chapter 1.

Theory and Practice of Sustainable Communities

As discussed in chapter 1, sustainability has many meanings, from an ecologically determined set of goals to principles by which community and industry activities may be assessed for their long-term impacts on society. It is not essential to agree on one definition of environmental sustainability, sustainable development, or sustainable communities to study the important transitions now taking place in public policy that are related to these issues. It is important, however, to understand the concept of sustainability, the social movements it has inspired, and especially the analytical challenges faced in measuring progress toward such goals. In chapter 2, Lamont Hempel surveys this terrain. He offers a thorough review of the different meanings attached to sustainability, the intellectual origins of the concept, and the competing orientations found today— from natural capital approaches to urban design, ecosystem management, and regional governance.

As Hempel's analysis suggests, building sustainable communities is a formidable challenge, whether in small towns or cities or in large metropolitan areas. In most locales, there will be substantial conflict over community goals—social, economic, and environmental—and it is not clear what form of public involvement or community processes will help most in developing consensus on those goals, or the public policies that can bring a community closer to them. There are both practical and theoretical issues raised by the many attempts to date to initiate this transition to sustainability.

At a practical level, hundreds of communities across the nation have initiated sustainable community projects that include "visioning" workshops and attempts to develop indicators or benchmarks for the core components of community sustainability: social well-being, economic vitality, and ecological integrity. Usually these endeavors are characterized by extensive participation and dialogue among stakeholders, sometimes assisted by outside consultants who specialize in community planning processes and systems modeling. Evidence of such citizen involvement can be found at the large number of World Wide Web sites on the subjects of sustainability and indicators, including several sponsored by federal agencies.[1]

Perusal of these sites indicates considerable potential for these local planning efforts. For example, selection of the most appropriate indicators (especially ones that are easily understood, up to date, and reliable, and encourage a long-term and integrated view of community sustainability) can help a community measure progress toward sustainability goals and also stimulate policy and behavioral changes supportive of them. Following the work of the President's Council on Sustainable Development, an Interagency Working Group on Sustainable Development Indicators has worked since 1994 on a framework for such indicators and specific recommendations (Farrell and Hart 1998).

Development of indicators and monitoring of environmental trends speak to another need we identify in chapter 1: to incorporate much more comprehensive and sophisticated data into environmental and economic planning and decision making—from ecological footprint analysis, to material and energy use, to computer modeling of human interaction with natural systems that can help inform policy choices. These techniques are also being used in selected locations already.[2] The need for such information and data management is particularly great at the local and regional level, where policy makers generally have had much less experience than their federal counterparts in using such technical information.

These recent activities also have implications for development of theory related to environmental decision making, public participation, and community planning, which we address later. Nearly every proposal of community sustainability incorporates ideas of intense civic engagement and cooperation among citizens in pursuit of long-term community goals, sometimes in concert with government officials and sometimes independent of them. As Robert Putnam (1993) has so persuasively argued, however, such ambitious collective action depends on the existence of sufficient "social capital," or the "norms and networks of civic engagement." To the extent that some communities and regions succeed in identifying shared social, economic, and environmental goals, and in fostering the development of a new generation of public policies grounded in sustainability, they merit close attention and analysis. The implications extend well beyond environmental policy.

Fostering Sustainability through Pilot Programs   These new kinds of community-level activities are precisely what the EPA has been trying to encourage through its Sustainable Development Challenge Grant Program (SDCG), one of many recent initiatives that the agency refers to collectively as Community-Based Environmental Protection (U.S. EPA 1997a). In this program, communities receive federal funds to leverage local public and private investments that aspire to integrate economic development, environmental protection, and social well-being through community partnerships, educational efforts, and voluntary action, in addition to conventional regulation.

A diversity of eligible activities fall under the SDCG program, including the use of integrated approaches to sustainable development; cleanup and redevelopment of contaminated property (brownfields); development of metropolitanwide transportation programs that promote sustainability; pollution prevention across multiple sectors of the community; efforts to reduce consumption, promote conservation, and minimize waste generation; and development of community awareness programs to publicize principles of stewardship, ecosystem management, and sustainability.

The challenge grant program has drawn great interest from environmental organizations and communities engaged in sustainability initiatives. Some 600 applications were received in the program's first year of operation in fiscal 1997, and about 960 applications in fiscal 1998. Unfortunately, in its most recent competition, the agency had funds to support only a very small fraction (about 4 percent) of those projects (U.S. EPA 1998).

A good example of this new EPA role is the establishment of a Sustainable Ecosystems and Communities Clearinghouse (SECC), operated by the Office of Sustainable Ecosystems and Communities (OSEC). OSEC provides information on local planning for sustainability through its Web site and other methods that are designed to build local capacity for problem solving (U.S. EPA 1996 and 1997a). For instance, the office funded development of a comprehensive citizens' guide to achieving sustainability, which was assembled by the Nature Conservancy for Virginia's Eastern Shore and its coastal ecosystem (Nature Conservancy 1996). It also has developed a training program for community leaders who are attempting to formulate sustainability indicators for their communities and regions (Farrell and Hart 1998).

Other offices within the EPA have used their financial resources, especially program grants, for similar ends, serving as catalysts for state and local environmental protection actions. Although modestly funded, such grants have stimulated innovative actions on pollution prevention, solid waste management, and wetlands protection; encouraged more effective water quality planning; promoted environmental justice; and supported environmental education (U.S. EPA 1997b).

Even without federal funding and technical support, however, cities as different as Chattanooga, Tennessee; Rochester, Minnesota; Green Bay, Wisconsin; Boulder, Colorado; Portland, Oregon; and Seattle, Washington, have taken action. They have established new voluntary, community-based planning processes to manage growth and promote economic development that is seen as compatible with area resources and priorities, sometimes with the assistance of new state goals and incentives (e.g., see Minnesota Planning 1995; Nature Conservancy 1996; Urban Ecology 1996; and Mazmanian et al. 1996).

Reinventing Environmental Regulation Related EPA efforts fall under the Clinton administration's regulatory reinvention program and reflect the recommendations of the vice president's National Performance Review of 1993. They have offered benefits similar to the state and local efforts described above, while illustrating the agency's determination to explore new approaches for addressing long-standing concerns about the cost, efficiency, intrusiveness, and flexibility of environmental protection policy. These include the Common Sense Initiative and Project XL, to promote greater flexibility, and the National Environmental Performance Partnership System (NEPPS), designed to improve working relations with the states and to strengthen the management, efficiency, and effectiveness of environmental programs. By 1998, well over one-half of the states had agreed to enter the partnership system (U.S. EPA 1997b; Kraft and Scheberle 1998).

New programs such as the NEPPS may well succeed in giving the most innovative and environmentally progressive states greater flexibility and in stimulating greater public involvement. The EPA can advance such initiatives, however, only within the limits set by Congress in the major environmental statutes, and where sufficient funds are provided through the annual appropriations process. As noted throughout this volume, these epoch one policies generally constrain the EPA and the states from moving too far from command-and-control approaches. Thus a transition to epochs two and three depends in good measure on congressional willingness to revise those statutes appropriately.

Research Needs   All of these sustainable community efforts are fairly new, and it is premature to judge their success at this time. Nonetheless, they indicate a substantial commitment on the part of the EPA's senior management, among other public officials, to break away from traditional command-and-control regulation. As is the case with regulatory reinvention in general, it is more difficult to determine how committed the EPA program staffs, state regulatory officials, and members of Congress and state legislators are to these initiatives, and the extent to which they are able and willing to supplement existing regulatory mandates with market-based approaches and support for sustainability activities.

Given the experimental and limited nature of most of these new endeavors, our knowledge of the paths to community sustainability could be enhanced through additional research. Of special value would be studies of both typical and unusual communities and assessments of the factors facilitating or constraining local initiatives and successful implementation of them—such as committed leadership, public support and involvement, long-range planning, and a linkage between sustainability goals and public policies that support them. Emphasis might also be given to the knowledge, attitudes, and behavior of key stakeholders, particularly their understanding of sustainability concepts such as systems thinking, and their willingness (or reluctance) to help further a transition to community sustainability.[3] Why are some community leaders willing to support or lead such initiatives and others not? What incentives might help to reduce the transaction costs involved?

In a similar vein, analysis of the effectiveness of government programs, such as the new ventures begun by the EPA on sustainable communities, and state and local efforts to promote sustainable planning initiatives would be invaluable for learning what works best. Which programs achieve their objectives and which do not, and what accounts for the difference? Where could any additional financial resources yield the greatest impact?

Much the same can be said of studies of public participation in community decision making. As noted earlier, advocates and analysts alike tend to make questionable assumptions about citizen interest in environmental issues and sustainability, which are often low-salience issues. Similarly, we tend to think about citizen involvement and community consensus building without full appreciation of their complexity. Yet these processes of public involvement and support are vital to both the second and third epochs of sustainable communities, especially the latter.

Thus we need to learn more about citizen capacity to play the roles expected in so many discussions of environmental decision making, and how that capacity might be enhanced at a time of widespread public distrust in formal institutions and in the political process (Nye, Zelikov, and King 1997). In light of these concerns, what are the best methods of public participation, and what can be done to help assure their success? Experience with public involvement in state comparative risk projects offers some insight (Minard 1996; Patterson and Andrews 1995), as do other studies of participation, including Williams and Matheny's (1995) creative analysis of democratic dialogue in community hazardous waste decisions.

J. Clarence Davies and colleagues in the Center for Risk Management at Resources for the Future have begun a major research project on public participation in environmental decision making that should help to build such knowledge. Among other topics, they are investigating innovative procedural, nongovernmental, and technological mechanisms for improving participation, based on an extensive examination of case studies in environmental decision making. They hope to show which forms of public involvement are most effective in achieving various social goals (McGovern and Beierle 1997). Citizen use of the Internet deserves special attention in any such investigation. Some 40 million of the nation's nearly 100 million households will have Internet access by 2001. What are the implications of this important development for citizen knowledge, learning, and involvement, especially in local sustainability efforts? Many similar questions regarding public participation in environmental policy and community sustainability decisions merit research and analysis.

Air and Water Pollution Control: The Appeal of Market Incentives and Collaborative Decision Making

The examples of community and EPA initiatives discussed above provide considerable support for the argument that a trend toward sustainable communities is increasingly evident in the United States, however inchoate and tentative the actions appear as of the late 1990s. What about the transition from the first to the second epoch that is the subject of chapters 3 and 4, particular as it applies to such conventional concerns as clean air and clean water? The cases we include here involve air pollution control in Los Angeles and water pollution control in Northeastern Wisconsin. Both illustrate the efforts by state, and especially local and regional, environmental agencies to adopt new approaches to pollution control to compensate for the inherent limitations of command-and-control regulation. They also demonstrate the appeal of those innovative tools to almost all policy actors today.

Overall, the epochs framework helps to both describe and explain the turmoil and changes in the air pollution arena in Los Angeles, from 1970 to 1990. The region fully embraced epoch one thinking and approaches. Then, in the late 1980s, it moved toward epoch two, efficiency-based regulatory restructuring in spirit. It continues in this vein. This new approach is reflected in the downsizing of the administrative and institutional apparatus of the Southern California Air Quality Management District, challenges from business and the community to the Federal Implementation Plan, loss of political support for the command-and-control approach, and a greater reliance on market-based approaches and new technologies (such as cleaner fuels, cars, and trucks) to achieve air quality goals. There has also been a greater sensitivity to the economic ramifications of clean air policies and programs. The district continues to move forward in its fundamental missions of air pollution reduction while changing from within its implementation philosophy, policy approaches, and the predominant players on the board and in the broader stakeholder community.

There is little evidence today, nor logical necessity, that the environmental movement of the region is moving forcefully into the third epoch of sustainability. Nonetheless, numerous communities in the metropolitan area, such as Pasadena, are beginning to develop sustainability goals and programs. At least some attention is being given to components of the third epoch, such as the need to address population growth in Southern California, to develop a limited mass transit system to relieve highway congestion and air pollution, and to attract new and less pollution industries to the area.

The case of water pollution control in Wisconsin offers several important lessons. Looking back, it is evident that a process of cleanup that began in the era of command-and-control in epoch one was eventually stymied by the combination of local opposition to costs and remedies imposed from a narrow view of options. Those conditions precipitated a search for alternative mechanisms of dispute resolution, which led eventually to the embracing of a strategy based on public-private partnerships and collaborative decision making. Moreover, in order to achieve the original water quality goals, public officials and community leaders were forced to address and include the broad and intrusive issues of nonpoint source water pollution (especially agricultural contributions), on the one hand, and to consider social and economic objectives of the community, on the other.

By 1997, it became apparent that this process was taking on the trappings of an epoch-three approach. It included comprehensive, multimedia environmental assessment, with extensive collaboration and partnerships involving environmental scientists, the business community, local and state officials, and concerned citizens. These efforts in turn stimulated new interest in the tools and approaches characteristic of epoch three. Public officials and community leaders are beginning to embrace the idea that community and regional planning and decision making, including growth management to help control urban sprawl, should be grounded in principles of sustainability and the use of comprehensive ecosystem or watershed management.

As the case study makes clear, however, there are also important limitations at this stage to what such partnerships and collaboration can achieve, especially when the costs of water pollution control and cleanup of contaminated river sediments are exceptionally high, and the evidence of societal benefits of cleanup uncertain. Cooperative decision-making processes are difficult to initiate and sustain, they can be enormously time-consuming and frustrating to all concerned, and they can easily fail if short-term economic concerns override long-term problem-solving strategies.

In the case of Northeastern Wisconsin, some of these constraints became readily apparent. The entry of the federal government under the Superfund Program dramatically altered the cast of policy actors, strategic considerations, and the financial stakes; it also served as a reminder of the continuing importance of epoch-one regulatory laws. Collaboration may work best in the early stages of community and regional planning when problems are being identified, research studies are commissioned, long-term goals are set, and consensus is relatively easy to achieve. This particular case also tells us that the force of federal law may still be needed to resolve disputes over the most appropriate cleanup standards, the specific methods and technologies to be employed, and who ultimately pays the bills.

As both case studies in this section of the book imply, new policy approaches such as collaborative decision making supplement, but do not entirely replace, the contribution made by command-and-control regulation. Similarly, devolution of responsibilities to the states offers important opportunities for innovation locally and regionally. Yet it also may encourage some states and communities to weaken environmental protection efforts in response to pressure from business interests and in reaction to short-term economic forces. Further research focusing on the conditions under which collaborative decision making, devolution, and similar approaches work best in achieving consensus and generating viable solutions, along with the factors that make them less successful, could help to guide their use in other venues. In-depth interviews with key policy actors could go a long way toward providing such information. As noted elsewhere in this chapter, attention to the diversity of communities and regions in the nation may help to clarify key variables affecting the use of new policy approaches, as well as the pace and character of a transition to sustainability.

Managing and Protecting the Land: Historical Patterns and New Directions

Whether at the local and state level or nationally, the way in which land resources are used is a central concern in present environmental policy and future sustainability. Yet the management of natural resources in many ways presents a different array of problems than does pollution control, and it has distinctive policy objectives and policy tools. Moreover, at least at the federal level, the management falls under the jurisdiction of agencies with only loose ties to the EPA—such as the U.S. Forest Service, National Park Service, Bureau of Land Management, and Fish and Wildlife Service. Nevertheless, many of the same transitions in public policy are evident in land use, as we have seen in pollution control, with new policy goals and fresh approaches increasingly evident (Kraft 1996; Davis 1997; Yaffee et al. 1996; Cortner and Moote 1998).

As Daniel Press argues in chapter 5, at the local level, we find increasing interest in setting aside land for parks, preserves, and recreation, to enhance quality of life and protect special areas for future generations. At the national level, sharp conflicts have arisen over the use of public lands, especially in the West. Traditional uses of public lands for commercial purposes such as mining, logging, and ranching increasingly have come under attack by environmentalists and public officials, like Interior Secretary Bruce Babbitt. They seek to curtail subsidies to traditional user groups and to protect a larger percentage of public lands under the banner of ecosystem management. At the state and local, as well as the national level, we can see considerable evidence of the transitions under way. Equally evident is that a diversity of social, economic, and political obstacles will constrain their success.

Many factors have contributed to the effort to protect agricultural and open-spaces in California over the course of the twentieth century (with similar developments evident in other states), extending back well before the first contemporary environmental epoch discussed in this book. The emergence of the first epoch did foster a rise in land acquisitions for protection and open-space use, through the preexisting methods of federal and state agency acquisitions, until the passage of Proposition 13 in 1978. In effect, the ambitions of acquisition agencies were heightened in this period, but the implementation philosophy, points of intervention, and institutional arrangements remained the same: a top-down, public agency approach.

With public funds from federal and state agencies largely choked off by the end of the 1970s, new methods, based on local initiatives often led by nonprofit land trusts, sprang up throughout the state and became the hallmark of the second environmental epoch with respect to land preservation. This process continued as the dominant form of land preservation policy into the 1990s, with the added consideration for wildlife and habitat protection reflecting the first appearance of a more comprehensive ecosystem-management and sustainability approach to land-use management, which is characteristic of epoch three.

At the federal level, one of the most striking trends has been a movement away from an emphasis throughout the nineteenth century and most of the twentieth century on economic development through resource extraction to a new posture of multiple use and, especially, ecosystem management. In particular, with adoption of new conservation-oriented public lands policies in the 1960s and 1970s (such as the Wilderness Act of 1964, the Endangered Species Act of 1973, the Federal Land Policy and Management Act of 1976, and the National Forest Management Act of 1976), the transition has become more pronounced. Environmentalists were successful in securing passage of the national legislation. Yet the new policy directions have drawn substantial opposition in many Western communities and states, and have sparked intense battles in the nation's capital (Switzer 1997; Cawley 1993).

It is easy to overlook the enormity of public lands managed by the federal government—30 percent of the nation's land surface—and the importance this has for moving to sustainability for the United States. With respect to the epochs approach, the transformations under way in public lands management suggest a movement from epochs one to three, bypassing in a number of important respects epoch two. That is, the epoch two characteristics—emphasis on the introduction of market mechanisms and a devolution of authority to states and localities—are less evident in public lands and other natural resource policies, even though much of the criticism of natural resource policies, as well as recommended new directions, emanate from these same ideas (Lowry 1999). Nevertheless, in formal congressional policy, as well as in agency actions and programs, the driving force in federal land management has been shifting from resource extraction to a model based on concepts of ecosystem-management practices (Clarke and McCool 1996; Cortner and Moote 1998). The extent to which past practices can be overcome and more sustainable, scientific, ecosystem-based management practices made workable and implemented remains a question for the nation as we move into the twenty-first century.

The assessments offered in this book on land use suggest the value of additional research on the forces shaping these transformations. As Press indicates, one way to study these developments is to compare systematically the communities that are implementing sustainable land use with those that are not, and try to isolate the key variables that account for the differences. Such knowledge adds invaluably to the more prescriptive accounts of sustainability. Too often such normative analysis tells us how essential it is for societies to embrace new norms of resource management based on ecological principles but offers little guidance concerning the social, cultural, economic, and political factors that affect such a fundamental shift in human behavior.

Similarly, any genuine progress toward sustainable communities in the West will depend on an understanding of the deeply held beliefs and values of residents of these communities, many of whom often bitterly oppose federal policy initiatives on ecosystem management. To what extent can these clashing values be reconciled and support be built for sustainable management of natural resources? What kinds of decision-making processes, public involvement mechanisms, education, and policy incentives offer the best hope for doing so? Are the prospects for building sustainable communities in the West improving with the decline of extractive industries, the rise of new business enterprises, and an influx of new residents?

Two quite different examples in the late 1990s indicate the potential for reconciling interests that historically have clashed on economic development and environmental protection. One involves success in the Sierra Nevada region along the California-Nevada border near Lake Tahoe in reorienting the local economy and fostering sustainability. A local Sierra Business Council worked to "secure and enhance the economic and environmental health of the region for future generations" (Christensen 1996). Among other activities, the group completed an audit of the natural, social, and financial capital of the region to enable better tracking of the area's well-being—from water quality and educational achievement levels to employment opportunities. Such an audit has long been encouraged by environmentalists, but the Sierra Business Council appears to be the first business group in the country to put such social cost accounting into practice in day-to-day decision making within both the public and private sector. Studies of how areas like this have succeeded in designing and adopting such new approaches would be enlightening. It is interesting as well that this subject has attracted the attention of major foundations, which may signal their interest in fostering similar efforts in other venues.

The other case, which received much attention and praise, involves land conservation near San Diego, California. In early 1997, a comprehensive, regional conservation plan developed under the Endangered Species Act was approved for a large expanse of land in an area rich in biodiversity. The plan, which some journalists referred to as "revolutionary," drew enthusiastic support from the mayor of San Diego and from environmentalists, developers, and other public officials. State and federal policy makers argue that management plans of this kind, integrating environmental and economic goals, could provide a model for the rest of the nation (Stevens 1997; Ayres 1997).

Studies of similar efforts in other regions and communities would be helpful for both the practical challenge of integrating economic and environmental concerns and the development of theory about the process of transition toward sustainable communities. What conditions have fostered the successes evident in these cases? For example, how important are local political leadership, the existence of persuasive economic and scientific analyses that confirm the long-term benefits of sustainability-based actions, or opportunities for dialogue among the various stakeholders?

Community and Regional Sustainability: Transformations in Progress
The last three chapters in the book, by Tugwell, McElwaine, and Fetting; Horan, Dittmar, and Jordan; and Rabe, share a common theme. Each describes new efforts to promote sustainable practices through comprehensive, multimedia strategies involving a multiplicity of actors, two within specific geographic areas (Pittsburgh and the Great Lakes Basin), and one emphasizing use of new technologies and new concepts in transportation, which apply nationally. The issues are different in each case, and yet all three chapters ask about the factors promoting the use of new approaches and those that constrain innovation and policy change.

Urban Revitalization and Sustainability in Pittsburgh Several lessons can be learned from the transformation to a more sustainable community under way in Pittsburgh—regarding root causes, contributing factors, and the importance of the broader environment and sustainability movement in helping to set the direction for the city. Few major U.S. cities have experienced either the depth of rust-belt industrial decline or the rapidity and breadth of a sustainability overlay on what had begun as a significant strategy of economic revival as has Pittsburgh. From thirty years of struggling under the pressures to clean up under epoch-one legislation, and barely being touched by epoch two's "new environmental strategies and institutions," Pittsburgh is beginning to chart its economic recovery in green terms as a matter of pragmatism in an environmental age. Public-private partnerships, local initiatives, seeking out green industries, along with marshaling the intellectual energy of regional universities, community spirit of non-for-profit organizations, and resources of leading regional foundations and opinion leaders are all part of the epoch-three blueprint. The incubation has begun, and the compass is set.

The success of the eco-economic renaissance in Pittsburgh—as forerunner, visionary, model of late twentieth century urban revival—if realized, will be an important benchmark in the sustainability movement. It is equally important to understand why all stakeholders have not embraced these new directions. As we have already argued, a fruitful line of research would be to ask about the conditions facilitating such an urban renaissance and those blocking it, and the perceptions, values, and attitudes of both supporters and opponents. What conditions might make pursuit of a greener Pittsburgh a more salient issue for the business community and political leaders? What incentives might attract their interest and support? Comparisons with developments in other cities would be instructive.

Transportation Policies and Community Sustainability   National transportation policy is clearly going through a transformation in thinking and action, a phenomenon vividly represented in the Intermodal Surface Transportation Efficiency Act of 1991. In some ways, like public lands policy, the transformation is from epochs one to three, with only minor forays into epoch-two kinds of activities, for example, public-private partnerships in building new freeway links in Southern California. The challenge of building sustainable—that is, integrated, comprehensive, and community-oriented—-transportation systems has been a subcurrent of national transportation planning for decades. Providing the statutory and financial mechanisms for doing so, however, awaited passage of ISTEA.

National transportation planning in the 1990s and beyond is now explicitly tied to consideration of air pollution control requirements (the Clean Air Act), quality of life considerations, and cross-modal (auto, bus, rail, plane) possibilities. The locus of decision making, moreover, has moved from Washington and state and regional transportation planning agencies to metropolitan planning organizations, where it is expected to be part of a more integrated metropolitan design. The extent to which the transformation in policy and formal institutional design is actually affecting project design and spending priorities remains and open question, as well as a central challenge, of the third environmental epoch.

The intensive partisan and regional battles that erupted over renewal of ISTEA during 1997 and 1998 provide some hints about how far the transformation in transportation planning has come and how far it has to go. In 1998, Congress overwhelmingly approved, and President Clinton signed, the Transportation Equity Act for the 21st Century, a bill that authorized more than $200 billion over six years to upgrade deteriorating highways and aging bridges and to expand the use of mass transit across the nation. Spending was increased by about 44 percent over that for ISTEA (with about 20 percent, or $41 billion, of the funds going to mass transit), making the 1998 measure the largest public works project in the nation's history. It should be said that members of Congress seemed to be far more interested in ensuring that their districts got a generous portion of highway funds than they were in the opportunity to foster sustainability through initiatives in transportation policy (Ota 1998). Nonetheless, new directions in transportation continue to be funded and assisted by the federal government, and analysts could profitably turn their attention to the factors that shape innovative transportation decision making in communities across the country. That is where the most important integration of transportation with other elements of sustainable communities will take place.

Regional Ecosystem Management: The Great Lakes Of all the cases of environmental policy transformation examined in this book, the thirty-year effort to reverse the contamination and ecological degradation of the Great Lakes is clearly the most ambitious, large-scale, complex, and multijurisdictional. In several ways, it is also the most informative in terms of history of efforts made, successful and not. Moreover, it is the most encouraging when viewed in light how much progress has been made, given the extraordinary difficulty of not only cleaning up the environment but also of evolving sustainable patterns of industrial production, agriculture, and human settlement. Whether one chooses to see the cup half-full or half-empty, it is at the halfway mark.

Barry Rabe's chapter on the Great Lakes illustrates the usefulness of dividing up the long history into more discrete epochs of environmental policy and focus. The command-and-control efforts at the outset have been replaced, for good or ill, by much greater local initiative, which relies far more on cooperation than on government compulsion. The necessity of a multimedia approach to protect the ecology of the lakes is not only acknowledged today, but it is also the cornerstone of water quality management strategy.

As is also evident in chapter 4, by Kraft and Johnson, there is no question that many elements of ecosystem management and long-term goals of sustainability have been widely embraced in the Great Lakes Basin. Despite those gains, however, many impediments remain. Further progress would seem to hinge on improved scientific knowledge of ecosystem functioning, increased public understanding of the issues, and stronger leadership from both the public and private sector. As much as the Great Lakes ecosystem has been studied, enough uncertainty remains to cast doubts on proposals to protect Great Lakes water quality, particularly when they involve controversial land-use decisions affecting nonpoint sources of pollution, and costly cleanup actions. Further research on how such conditions affect local, state, and provincial actions promoting sustainability would be useful. As many areas within the basin embrace the goal of sustainability, it becomes even more crucial to understand why others do not attach as high a priority to such actions, and what incentives or other stimuli might push them to change direction.

Transitions and Transformations Revisited

It is important as we conclude to acknowledge once again the impressive gains the United States has made, especially in the past three decades, in controlling air, water, and land pollution and protecting the nation's precious natural resources (Council on Environmental Quality 1997). Dozens of major environmental laws have been enacted and implemented at the federal level, and at all levels of government a vast array of significant changes have been instituted in the way we make decisions that affect the environment. Environmentalists and government officials can justifiably claim credit for having moved the nation toward a greener future.

As many of the chapters in this volume make clear, however, the federal policies (and, we should add, many of their state counterparts) that have helped to produce these improvements in environmental quality are deeply flawed and unlikely to promote a transition to sustainable communities. This is an assessment increasingly shared by analysts and decision makers from a diversity of disciplinary backgrounds and policy orientations (e.g., Davies and Mazurek 1998; Mazmanian and Morell 1992; National Commission on the Environment 1993; Vig and Kraft 1997). At the same time, experimentation is under way at all levels of government as well as within the private sector to shift directions and try new approaches (Sexton et al. 1999). Such innovation is particularly evident—and promising—at the state and local level (National Academy of Public Administration 1994; John 1994; Hamilton 1990; Press and

Mazmanian 1997; Rabe 1997; Hockenstein, Stavins, and Whitehead 1997; Kraft and Scheberle 1998; Mazmanian, et al. 1996).

Such varied experiments in widely different venues reveal a common concern for building robust, flexible, and adaptive environmental policies for the twenty-first century. Along with many others, we believe this transformation in policy design will be most effective if anchored in the concept of sustainability (President's Council on Sustainable Development 1996; Sitarz 1998; United Nations 1992; U.S. EPA 1997b; Hempel 1996; Trzyna 1995). Such an approach could provide a much more comprehensive and holistic framework for integration of economic and environmental policies, and for long-term ecosystem management and protection of public health that is largely beyond the capacity of present environmental protection and natural resource statutes.[4] Of necessity, the kind of transformations that we believe are needed will not come quickly, and they will not be equally evident in the extraordinarily diverse set of communities within the United States—and globally.

Several of the chapter authors also remind us that these processes of social, institutional, and policy change need to be carefully assessed. We ought to be asking about the likely effects of new policy approaches, short term and long term, both positive and negative. Where the data exist, it would be useful as well to document actual impacts on environmental outcomes and other sustainability goals. For example, before abandoning regulatory policies that have been moderately effective, if not always efficient, we ought to have some degree of confidence that improvements in environmental quality will follow. Where the evidence indicates that new approaches are indeed producing better results (and are likely to do so in the future), we also want to know why. What helped to bring about the desired behavioral and institutional changes? Such knowledge would be important for further policy design efforts (Sabatier and Jenkins-Smith 1993; Schneider and Ingram 1997). Inquiry of this kind, borrowing from the fields of policy analysis and program evaluation, is beginning to attract academic and practitioner interest (Fischer and Black 1995; Chertow and Esty 1997; Knaap and Kim 1998; Ringquist 1995; Stavins 1991; Sexton et al. 1999; Barnthouse et al. 1998).

Final Observations

We hope the analyses and conclusions we have summarized here stimulate others to revise and extend the work represented in this volume. However insightful it may be, research based on case studies and analytic narratives is inherently limited in terms of generalization to other cases, with their varying timetables and with features unique to time, place, and focus. The six case studies in the book are illustrative, we think, of the problems faced in hundreds of other communities and regions across the country, and of the challenges facing citizens and policy makers as they seek to reform environmental policy and reconcile it with economic development and other social needs. We believe the cases are sufficiently suggestive of the existence of the transformations under way that they facilitate understanding, and further inquiry and involvement in these remarkable experiments in democracy.

For the environmental policy arenas developed most fully during the first epoch—that is, air, water, noise, and toxic wastes that focused on specific media—the transformation from epoch one to epoch two has been taking place across the board of policy domains. In these major areas of explicit environmental protection policy, the framework of the three epochs finds the strongest application.

In some instances, such as with water pollution in the Great Lakes, the evolution across the epochs has been more evolutionary—under way, in concept at least, from the outset and linking relatively easily to the ideas being put forth today—and guided by some of the key principles of sustainability. Realization of these principles, of course, has not been easy in the Great Lakes, and practice does not always following theory.

In the area of land-use policy, management and planning go back much further than the 1970s, and are related to many more considerations than environmental protection. If anything, modern-day land-use practices are being affected by the sustainability movement, but it is not clear that practices are keeping apace. Quite simply, managing the land in the United States is more difficult than managing for clear air, clean water, or other media-specific resources.

Transportation policy, while fitting the epochs framework somewhat, only recently has been held accountable to environmental considerations, the requirement under the National Environmental Policy Act for environmental impact statements notwithstanding. That is, the design and funding of projects supported from ISTEA are today supposed to be within a broad sustainable communities framework.

What clearly has changed from the first to the second, and now to the third epoch, is the points of intervention, policy tools, and institutional venues. The institutional locus of decision making has been moving toward the state and local level, with substantially greater local determination.

Complementing this devolution of responsibilities has been far greater use of collaboration and partnerships of all kinds. In addition, command-and-control regulation has slowly been giving way to new approaches grounded in market-based concerns and incentives. Enforcement today may be top-down, with the EPA continuing to play a strong role. But policy deliberations, rule making, and the use of market-based solutions are increasingly being defused throughout society, in particular to locally affected stakeholders.

In terms of information and management needs, there is no question that the demand for good scientific data is growing by leaps and bounds, as is the quantity of data being produced by technical personnel both in and outside government. The most difficult challenge for today is not simply to acquire more data, but how to link the data scientists and monitoring systems can provide with prudent resource management decisions. This requires creative solutions for bringing together scientists, agency managers, citizens, and other stakeholders in a way that facilitates assessment of the data, however limited at any given time, and fosters a public discourse over implications for public policy and community well-being. We have a long way to go to find the best way to achieve these goals.

One of the important messages to come out of the collection of specific, real-world case studies is the extent to which the processes and trends key to the transformations under way in the environmental movement have a parallel in other fields of inquiry. Our epochs model is very much akin to the idea of punctuated equilibrium models in public policy (Baumgartner and Jones 1993). The long-standing debate between the efficacy of top-down and bottom-up policy implementation, discussed by

Rabe, has been tested empirically over time in the environmental arena, and it appears that some version of bottom-up, or at least bottom-level involvement and collaboration, is more efficacious.

The cases included here illustrate, on the one hand, that no simple analytical framework can do full justice to the realities of environmental policy and life as it actually unfolds. On the other hand, understanding the broader contours of the many changes now occurring in environmental policy and practice, in terms of more encompassing epochs, is a genuine aid in helping to see basic dimensions and directions of change.

The analytical framework we offer in this volume provides a useful way of thinking about these changes in the environmental movement. It can be applied in broad, sweeping terms—moving from the first through the third epoch as a way of providing an historical sweep and narrative. It identifies the full range of particular dimensions of epoch-level change, from ideas through implementation strategy. As presented here, the framework emerged out of the transformations under way in the air pollution arena—so dominant in the United States beginning in the early 1970s. It appeared to fit that arena well in both general and specific terms.

While the general transformations are equally applicable in each of the other arenas examined, the timing and close fit of detailed changes are less applicable as we move from air to watershed, open-space, and public lands policies. The closeness of fit in particular aspects of change is even less apt in viewing the more multimedia arenas that are brought into consideration in national transportation, urban revitalization, and supraregional, transnational arenas, such as the Great Lakes. The effects of epoch one are evident in each of these arenas, as well as the emergence of epoch three. However, epoch two is far less clearly demarcated in these cases.

Inquiry into other cases and other venues will surely help refine the framework and broaden our understanding of the dynamics of environmental policy as we enter the twenty-first century. We believe that such research can also provide much needed strategic advice to planners, analysts, policy makers, and citizens who will have to choose and live with the policies that govern environmental quality over the next several decades, when the nation and the world must come to terms with the challenge of environmental sustainability.

Notes

1. Among the most useful are: the Department of Energy's site (www.sustainable.doe.gov); the Global Environmental Options site, which has some 400 sustainable development links (www.geonetwork.org/links); Hart Environmental Data (www.subjectmatters.com/indicators); and the Center for Sustainable Communities Hotlinks (weber.u.washington.edu/~common/hotlinks.html).

2. One example of such computer modeling was used in a workshop in Green Bay, Wisconsin, in April 1997, and plans are being made for much more extensive use in the future. The modeling is derived from existing systems models, such as Stella, and it is used to scope a problem and set out issues for discussion. The advantage of using computer simulations is that they can be programmed to incorporate dynamic relationships among the major environmental, demographic, and economic variables that are otherwise difficult to understand and consider. Such models could greatly assist business and community leaders and government officials in identifying appropriate paths of development and in facilitating discussion of their implications for community quality of life, the environment, and the economy. Participants can experiment directly with the model, running simulations to illustrate the consequences of different kinds of decisions about community development. A suitable forum could be established to discuss the implications of the exercise with a broader group of stakeholders.

3. In one of the few empirical studies of these questions, Mark Lubell and colleagues have investigated the environmental, economic, and political conditions under which ecosystem partnerships at the local level are likely to emerge. Such partnerships at the ecosystem level (e.g., watersheds) are intended to address common environmental problems through cooperation among stakeholders. In an early report (Lubell et al. 1998), they indicate that such partnerships are more likely to form when the "potential Pareto-benefits of cooperation outweigh the transaction costs involved for key participants."

4. The President's Council on Sustainable Development (PCSD) sponsored a National Town Meeting for a Sustainable America in May 1999, in the hope of catalyzing a national movement toward sustainable development. The PCSD brought together representatives from government, business, community, and nonprofit organizations to showcase ideas, technologies, and practices that reflect an integrated approach to economic, social, and environmental goals. It hoped to galvanize further action to implement sustainable development throughout the nation. See the PCSD's Web page: www.whitehouse.gov/pcsd/ntm.html.

References

Ayres, B. Drummond, Jr. 1997. "San Diego Backs 'Model' Nature-Habitat Plan." *New York Times,* March 30, A10.

Barnthouse, Lawrence W., Gregory Biddinger, William Cooper, James Fava, James Gillette, Majorie Holland, and Terry Yosie, eds. 1998. *Sustainable Envi-*

ronmental Management. Pensacola, FL: Society of Environmental Toxicology and Chemistry, Report from the Sustainability-Based Environmental Management Workshop, Pellston, Michigan, August 1993.

Baumgartner, Frank R., and Bryan D. Jones. 1993. *Agendas and Instability in American Politics.* Chicago: University of Chicago Press.

Cawley, R. McGreggor. 1993. *Federal Land, Western Anger: The Sagebrush Rebellion and Environmental Politics.* Lawrence: University Press of Kansas.

Chertow, Marian R., and Daniel C. Esty, eds. 1997. *Thinking Ecologically: The Next Generation of Environmental Policy.* New Haven, CT: Yale University Press.

Christensen, Jon. 1996. "High-Country Common Ground: In the Sierras, Growth and Preservation Are Not at Odds." *New York Times,* November 30, 21–22.

Clarke, Jeanne Nienaber, and Daniel C. McCool. 1996. *Staking Out the Terrain: Power and Performance Among Natural Resource Agencies,* 2nd ed. Albany: State University of New York Press.

Cortner, Hanna J., and Margaret A. Moote. 1998. The *Politics of Ecosystem Management.* Washington, DC: Island Press.

Council on Environmental Quality. 1997. *Environmental Quality: 25th Anniversary Report of the Council on Environmental Quality.* Washington, DC: Government Printing Office.

Davies, J. Clarence, ed. 1996. *Comparing Environmental Risks: Tools for Setting Government Priorities.* Washington, DC: Resources for the Future.

Davies, J. Clarence, and Jan Mazurek. 1998. *Pollution Control in the United States: Evaluating the System.* Washington, DC: Resources for the Future.

Davis, Charles, ed. 1997. *Western Public Lands and Environmental Politics.* Boulder, CO: Westview Press.

Farrell, Alex, and Maureen Hart. 1998. "What Does Sustainability Really Mean? The Search for Useful Indicators." *Environment* 40 (November):4–9, 26–31.

Fischer, Frank, and Michael Black, eds. 1995. *Greening Environmental Policy: The Politics of a Sustainable Future.* New York: St. Martin's.

Hamilton, Michael S., ed. 1990. *Regulatory Federalism, Natural Resources and Environmental Management.* Washington, DC: National Academy of Public Administration.

Hempel, Lamont. 1996. *Environmental Governance: The Global Challenge.* Washington, DC: Island Press.

Hockenstein, Jeremy B., Robert N. Stavins, and Bradley W. Whitehead. 1997. "Crafting the Next Generation of Market-Based Environmental Tools." *Environment* 39 (May):13–20, 30–33.

John, DeWitt. 1994. *Civic Environmentalism: Alternatives to Regulation in States and Communities.* Washington, DC: CQ Press.

Knaap, Gerrit J., and Tschangho John Kim, eds. 1998. *Environmental Program Evaluation: A Primer.* Champaign, IL: University of Illinois Press.

Kraft, Michael, E. 1996. *Environmental Policy and Politics: Toward the Twenty-First Century.* New York: HarperCollins.

Kraft, Michael E., and Denise Scheberle. 1998. "Environmental Federalism at Decade's End: New Approaches and Strategies." *Publius: The Journal of Federalism* 28:1 (Winter 1998), 131–146.

Lowry, William R. 1999. "Natural Resource Policies in the Twenty-first Century." In *Environmental Policy: New Directions for the Twenty-first Century*, eds. Norman J. Vig and Michael E. Kraft. Washington, DC: CQ Press.

Lubell, Mark, Mihriye Mete, Mark Schneider, and John Scholz. 1998. "Cooperation, Transaction Costs, and the Emergence of Ecosystem Partnerships." Paper presented at the annual meeting of the American Political Science Association. Boston, September 3–6.

Mazmanian, Daniel, Lamont Hempel, Thomas Horan, and Dan Jordan. 1996. "Sustainable Communities: The U.S. Experience." Paper prepared for the Berlin Conference on Sustainable Development, March 19–21.

Mazmanian, Daniel, and David Morell. 1992. *Beyond Superfailure: America's Toxics Policy for the 1990s.* Boulder, CO: Westview Press.

McGovern, Michael H., and Thomas C. Beierle. 1997. "E-Part: The Future of Public Involvement?" *Center For Risk Management Newsletter* (Resources for the Future), No. 12 (Fall):1–3.

Milbrath, Lester W. 1989. *Envisioning a Sustainable Society: Learning Our Way Out.* Albany: State University of New York Press.

Minard, Richard A., Jr. 1996. "CRA and the States: History, Politics, and Results." In *Comparing Environmental Risks*, ed. Davies. Pp. 23–61.

Minnesota Planning. 1995. "Common Ground: Achieving Sustainable Communities in Minnesota." St. Paul, MN: Minnesota Planning.

National Academy of Public Administration. 1995. *Setting Priorities, Getting Results: A New Direction for EPA.* Washington, DC: National Academy of Public Administration.

National Academy of Public Administration. 1994. *The Environment Goes to Market: The Implementation of Economic Incentives for Pollution Control.* Washington, DC: National Academy of Public Administration.

National Commission on the Environment. 1993. *Choosing a Sustainable Future: The Report of the National Commission on the Environment.* Washington, DC: Island Press.

Nature Conservancy. 1996. *A Citizen's Guide to Achieving a Healthy Community, Economy, and Environment.* Leesburg, VA: Center for Compatible Economic Development.

Nye, Joseph, Philip D. Zelikow, and David C. King, eds., 1997. *Why People Don't Trust Government.* Cambridge, MA: Harvard University Press.

Ota, Alan K. 1998. "American Hits the Highways, and Congress Must Navigate." *CQ Weekly*, May 16, 1266–1272.

Patterson, Christopher J., and Richard N.L. Andrews. 1995. "Procedural and Substantive Fairness in Risk Decisions: Comparative Risk Assessment Procedures." *Policy Studies Journal* 23 (Spring):85–95.

President's Council on Sustainable Development. 1996. *Sustainable America: A New Consensus for Prosperity, Opportunity, and a Healthy Environment for the Future.* Washington, DC: President's Council on Sustainable Development, February.

Press, Daniel, and Daniel Mazmanian. 1997. "The Greening of Industry: Achievement and Potential." In *Environmental Policy in the 1990s,* eds. Vig and Kraft. Pp. 255–277.

Putnam, Robert. 1993. *Making Democracy Work: Civic Traditions in Modern Italy.* Princeton: Princeton University Press.

Rabe, Barry. 1997. "Power to the States: The Promise and Pitfalls of Decentralization." In *Environmental Policy in the 1990s,* eds. Vig and Kraft. Pp. 31–52.

Ringquist, Evan J. 1995. "Evaluating Environmental Policy Outcomes." In *Environmental Politics and Policy,* 2nd ed., edited by James P. Lester. Durham: Duke University Press. Pp. 303–327.

Sabatier, Paul A., and Hank C. Jenkins-Smith. 1993. *Policy Change and Learning: An Advocacy Coalition Approach.* Boulder, CO: Westview Press.

Schneider, Anne Larason, and Helen Ingram. 1997. *Policy Design for Democracy.* Lawrence, KS: University of Kansas Press.

Sexton, Ken, Alfred A. Marcus, K. William Easter, and Timothy D. Burkhardt, eds. 1999. *Better Environmental Decisions: Strategies for Governments, Businesses, and Communities.* Washington, DC: Island Press.

Sitarz, Daniel. 1998. *Sustainable America: America's Environment, Economy and Society in the 21st Century.* Carbondale, IL: EarthPress.

Stavins, Robert. 1991. "Project 88—Round II, Incentives for Action: Designing Market-Based Environmental Strategies." Washington, DC: A public policy study sponsored by Senator Timothy E. Wirth and Senator John Heinz, May.

Stevens, William K. 1997. "Disputed Conservation Plan Could Be Model for Nation." *New York Times,* February 16, 8.

Switzer, Jacqueline Vaughn. 1997. *Green Backlash: The History and Politics of Environmental Opposition in the U.S.* Boulder, CO: Lynne Rienner.

Trzyna, Thaddeous, ed. 1995. *A Sustainable World: Defining and Measuring Sustainable Development.* Sacramento, CA: California Institute of Public Affairs.

United Nations. 1992. *Agenda 21: The United Nations Programme of Action from Rio.* New York: United Nations.

U.S. Environmental Protection Agency. 1998. "Sustainable Development Challenge Grant Proposal Guidance 1998." Washington, DC: EPA Office of Air and Radiation, August.

U.S. Environmental Protection Agency. 1997b. *EPA Strategic Plan.* Washington, DC: Office of the Chief Financial Officer, EPA 190-R-97-002, September.

U.S. Environmental Protection Agency. 1997a. *People, Places, and Partnerships: A Progress Report on Community-Based Environmental Protection.* Washington, DC: Office of the Administrator, EPA-100-R-97-003, July.

U.S. Environmental Protection Agency. 1997b. *EPA Strategic Plan.* Washington, DC: Office of the Chief Financial Officer, EPA 190-R-97-002, September.

U.S. Environmental Protection Agency. 1996. "Information Packet for OSEC." Washington, DC: Office of Sustainable Ecosystems and Communities, December.

Urban Ecology. 1996. *Blueprint for a Sustainable Bay Area.* San Francisco: Urban Ecology, Inc., November.

Vig, Norman, and Michael Kraft, eds. 1997. *Environmental Policy in the 1990s: Reform or Reaction?*, 3rd ed. Washington, DC: CQ Press.

Vig, Norman, and Michael Kraft, eds. 1984. *Environmental Policy in the 1980s: Reagan's New Agenda.* Washington, DC: CQ Press.

Williams, Bruce A., and Albert R. Matheny. 1995. *Democracy, Dialogue, and Environmental Disputes: The Contested Language of Social Regulation.* New Haven, CT: Yale University Press.

Yaffee, Steven L., Ali F. Phillips, Irene C. Frentz, Paul W. Hardy, Sussanne M. Maleki, and Barbara E. Thorpe. 1996. *Ecosystem Management in the United States: An Assessment of Current Experience.* Washington, DC: Island Press.

Index